Statistical Quality Control for the Food Industry

Second Edition

Statistical Quality Control for the Food Industry

Second Edition

Merton R. Hubbard

Consultant, Hillsborough, California

CHAPMAN & HALL

An International Thomson Publishing Company

New York • Albany • Bonn • Boston • Cincinnati • Detroit • London • Madrid • Melbourne
Mexico City • Pacific Grove • Paris • San Francisco • Singapore • Tokyo • Toronto • Washington

Cover Design: Andrea Meyer, emDASH inc.

Copyright © 1996
Chapman & Hall

Printed in the United States of America

For more information, contact:

Chapman & Hall
115 Fifth Avenue
New York, NY 10003

Thomas Nelson Australia
102 Dodds Street
South Melbourne, 3205
Victoria, Australia

Nelson Canada
1120 Birchmount Road
Scarborough, Ontario
Canada, M1K 5G4

International Thomson Editores
Campos Eliseos 385, Piso 7
Col. Polanco
11560 Mexico D.F.
Mexico

Chapman & Hall
2-6 Boundary Row
London SE1 8HN
England

Chapman & Hall GmbH
Postfach 100 263
D-69442 Weinheim
Germany

International Thomson Publishing Asia
221 Henderson Road #05-10
Henderson Building
Singapore 0315

International Thomson Publishing-Japan
Hirakawacho-cho Kyowa Building, 3F
1-2-1 Hirakawacho-cho
Chiyoda-ku, 102 Tokyo
Japan

1 2 3 4 5 6 7 8 9 10 XXX 01 00 99 97 96

Library of Congress Cataloging-in-Publication Data

Hubbard, Merton R.
　　Statistical quality control for the food industry / Merton R. Hubbard. —2nd ed.
　　　　p.　　cm.
　　Includes bibliographical references and index.
　　ISBN 0–412–07661-6
　　1. Food industry and trade—Quality control.　I. Title.
TP372.6.H83　1996
664'.07—dc20　　　　　　　　　　　　　　　　　94–44229
　　　　　　　　　　　　　　　　　　　　　　　　　CIP

Visit Chapman & Hall on the Internet: http://www.chaphall.com/chaphall.html

British Library Cataloguing in Publication Data available

To order this or any other Chapman & Hall book, please contact **International Thomson Publishing, 7625 Empire Drive, Florence, KY 41042.** Phone: (606) 525-6600 or 1-800-842-3636.
Fax: (606) 525-7778, e-mail: order@chaphall.com.

For a complete listing of Chapman & Hall's titles, send your requests to
Chapman & Hall, Dept. BC, 115 Fifth Avenue, New York, NY 10003.

Contents

Preface to the Second Edition

Within the six years since the first edition was published, ISO 9000, HACCP, Expert Systems, six-sigma, proprietary vendor certification programs, sophisticated team techniques, downsizing, new electronic and biochemical laboratory methods, benchmarking, computer-integrated management, and other techniques, standards and procedures descended upon the quality control managers in the food industry with the impact of a series of tornados. Everything changed; it was time to rewrite *Statistical Quality Control for the Food Industry.*

Or so it seemed. But, as it turns out, everything has *not* changed. The concepts of variability, sampling, and probability are still the same. The seven basic tools of statistical quality control still work. Control charts still supply the information to control the process—although now the computer is doing most of the calculations and graph construction faster, and in color.

On close examination, even some of the major developments are not really all that new. For example, ISO 9000 closely resembles *Food Processing Industry Quality System Guidelines* published in 1986, and some other quality systems. The powerful Hazard Analysis Critical Control Point technique has also been around for some time, and many food companies have been using selected portions of it voluntarily. Now, however, it has become part of the food laws and has suddenly received widespread publicity.

There *have* been some real changes, however. The power of the computer has been applied to several phases of the food industry: Management has found that some computer applications can reduce the need for manpower. Other computers have been harnessed to processes to receive electrical information, analyze the input, and instantly send adjustment signals back—thus improving the process by reducing variability. Some have been used to instantly provide expert process advice to the operator. Still others have been used to extract data from process computers, analyze, calculate, and produce graphs, charts, and reports for product and process improvement studies for immediate use by all interested departments.

Considering the ability of food processing companies to consistently manufacture safe foods with uniform quality over the past 20 or 30 years *without* these
new tools and new systems, one might expect that quality control improvements
would be marginal. On the other hand, these changes have already provided substantial opportunities for process and product improvement. This second edition
is intended to update the basic concepts and discuss some of the new ones.

Preface to the First Edition

If an automobile tire leaks or an electric light switch fails, if we are short-changed at a department store or erroneously billed for phone calls not made, if a plane departure is delayed due to a mechanical failure—these are rather ordinary annoyances which we have come to accept as normal occurrences.

Contrast this with failure of a food product. If foreign matter is found in a food, if a product is discolored or crushed, if illness or discomfort occurs when a food product is eaten—the consumer reacts with anger, fear, and sometimes mass hysteria. The offending product is often returned to the seller, or a disgruntled letter is written to the manufacturer. In an extreme case, an expensive law suit may be filed against the company. The reaction is almost as severe if the failure is a difficult-to-open package or a leaking container. *There is no tolerance for failure of food products.*

Dozens of books on quality written for hardware or service industries discuss failure rates, product reliability, serviceability, maintainability, warranty, and repair. Manufacturers in the food industry cannot use these measurements: food reliability must be 100%, failure rate 0%. Serviceability, maintainability, warranty, and repair are meaningless terms to food processors.

Consequently, this book on food quality does not concern itself with reliability and safety. It is assumed that manufacturers in the food industry recognize the intolerance of customers and the need, therefore, for rigid requirements for producing 100 percent safe and reliable product. Those few food processors who experience off-flavor, foreign material, salmonella, botulism, or other serious defects in their products rarely survive.

This book covers the various techniques for assuring the continuous safe production of uniform foods. All discussions are specifically tied to food industry applications. The chapter on statistics fundamentals is made palatable by using examples taken directly from companies processing fruits, wine, nuts, and frozen foods. Other food product examples are used to illustrate the procedures for generating control charts.

By now, most upper managers are aware that process control is a technique which long ago supplanted the inspect-and-sort concept of quality control. This book is intended to present these managers with an understanding of what the technique includes. It is also targeted at quality engineers, managers, and technicians who have been unable to find workable explanations for some of the quality techniques specifically used by the food industry. A new audience for this subject includes all departments in companies embracing the concept of total quality control. Here is a collection of quality techniques which accounting, procurement, distribution, production, marketing, and purchasing can apply to their departments. Finally, the book is aimed at students hoping to enter the field of food quality control and technicians aspiring to management positions in quality control.

Guidelines for overall quality control systems and suggestions for implementing a quality control program are discussed from a generic point of view; all other subjects are discussed in a very specific, "how to" manner. For example, an entire chapter is devoted to a step-by-step procedure for controlling the net quantity of packaged foods. It explains how to obtain data, interpret government weight regulations, calculate both legal and economic performance, and set target weights. For the most part, the calculations have been reduced to simple arithmetic.

Whenever possible, chapters have been designed to stand alone. As an example, the chapter on process control explains how charts are interpreted and what actions should be taken. While reading this chapter, it is not necessary to thumb through the pages to consult the Appendix or the chapter on methods for preparing control charts. Similarly, the chapter on design of experiments uses some concepts introduced earlier, but does not require the reader to review the chapter on fundamentals. Experiment design is a complex subject, but the book reduces it to a straightforward explanation and provides food processing examples, as well as a series of diagrams of the most useful designs.

The bibliography contains most of the common texts on statistical process control. In addition, the chapter on test methods provides a list of references which have food industry applications. The Appendix includes only tables referred to in this book.

The author has attempted to avoid theories and generalities in order to make this book as practical and useful as possible. In the immense field of food processing, it is remarkable how little specific quality control information has been available. It is hoped that this book will fill that gap.

Acknowledgments

The need for a book of this scope became apparent during the annual presentations of Statistical Process Control Courses for the Food Industry, sponsored by the University of California at Davis. I am primarily grateful to Robert C. Pearl, who has spearheaded these quality control courses since the early 1960s, and to Jim Lapsley, his successor, for continuing support during the book's development.

Many university staff and quality professionals have contributed to the preparation and instruction of these courses, and I must give special thanks to the following for permission to include portions of their unpublished notes in various sections of the book:

Professor Edward Roessler	Wendell Kerr
Dr. Alan P. Fenech	Chip Kloos
Sidney Daniel	Ralph Leporiere
Ronnie L. De La Cruz	Jon Libby
Seth Goldsmith	Donald L. Paul
Randy Hamlin	Sidney Pearce
Gilbert F. Hilleary	Floyd E. Weymouth
Mary W. Kamm	Tom White

I am grateful to the Literary Executor of the late Sir Ronald A. Fisher, F.R.S., to Dr. Frank Yates, F.R.S. and the Longman Group Ltd., London for permission to reprint Table XXXIII from their book *Statistical Tables for Biological, Agricultural and Medical Research* (6th Edition, 1974).

Thanks also to my wife Elaine for her professional help as my editor and for her encouragement and patience over the long haul.

Statistical Quality Control for the Food Industry
Second Edition

1 Introduction

In 1989, the United States imported $160 billion worth of manufactured products from the Orient. Prices for these products were not necessarily lower than for those produced in the United States, but the quality level and uniformity were excellent—a far cry from the reputation for shoddy goods that the Orient suffered throughout the first half of the twentieth century. This has raised fears that the Orient would ultimately take over the production of all of our products and that the United States has already turned into a service industry nation.

Statisticians have been known to generate analyses that are mathematically correct and yet occasionally open to question if the data are presented out of context. The $160 billion may fall into that category. There is no question that it represents a sizeable quantity of goods, but the yearly U.S. Gross National Product (GNP) in the late 1980s was $4 trillion. In other words, imports from the Orient accounted for less than 4% of our GNP. Four percent is not a dangerously high level—certainly not high enough to suggest that we are in danger of having all of our products manufactured elsewhere.

Government statisticians have replaced Gross National Product (GNP) with Gross Domestic Product (GDP), but there is only a small difference between these figures. Perhaps the 4% figure might be overly pessimistic, because trade imbalances have built-in correcting devices. For example, during periods when domestic consumption slows, imports will slow as well. United States exports during the end of the century were actually growing at a 25% annual rate, and trade deficits with foreign countries had peaked. This improvement was masked by the fluctuating value of the dollar against foreign currencies (Merrill Lynch, *Global Investment Strategy,* 21 March 1995; also the Japan Business Information Center, Keizai Koho Center).

The quality control and quality level of U.S. manufacturers are not necessarily inferior, as is implied by cold numbers alone. Perhaps the impression that our manufacturing industry is about to be taken over is due to the Orient's selection of some highly visible consumer products. Their manufacturers have done an excellent job in the photography, optics, automotive, and electronics fields. Even in these areas, however, the United States has a significant number of profitable operations with notable market share.

Food production and processing in the United States are areas of outstanding quality unmatched by the Orient. The most obvious example of food quality control is the safety of foods in the United States. There are 250 million people in this country who eat a total of about 750 million meals a day, or 274 billion meals annually. A benchmark study by the Centers for Disease Control (1981) analyzed the numbers and causes for foodborne disease outbreaks across the country for an entire year. They found 460 reported outbreaks of food poisoning, with an outbreak defined as two or more people becoming ill from the same food eaten at the same time. This figure represents the cases reported by public health departments, doctors, and hospitals, but does not include people who went to their own physicians or did not seek treatment. Such data are unavailable.

Working only with the proven data, the 460 figure, expressed as percent product failure, indicates that

$$(460 \times 100) / 274,000,000,000 = 0.00000017\%$$

It is difficult to find another country which has achieved that kind of record for any product in any industry.

If the record for food quality is so superb, why this book? Because the need to improve quality control is unending. Safety may appear adequate, but improvements are always possible in net weight control, product color, flavor, keeping qualities, production cost, absence of defects, productivity, and so on. There is little factual quality information provided at the graduate level of business schools, and much of the information available to the management of food processing companies is supplied by the quality control professionals who are likely to read this book. Perhaps management may find the time to read it as well.

VARIABILITY

We live in a world of variability. The person who first used the expression "like two peas in a pod" was not looking very carefully. No two people are exactly alike—not even twins. Astronomers tell us that, in this vast universe, no two planets are alike. Two man-made products that are "within specifications" may seem to be the same, but on closer inspection, we find that they are not identical.

Perfection is not possible. You know it, your friends know it, your children know it, but the chief executive officer of many companies does not admit to it. He says that there is no reason why all products in a properly maintained production line—with adequately trained and motivated workers, the right raw materials, expert supervisors, and quality control employees who know what they are doing—cannot be perfectly uniform, with no defects and no variation.

Although we must accept the fact that variability does exist, there are methods to control it within bounds which will satisfy even the chief executive officer. You shall find that:

- Statistical tools are available.
- Processes can be controlled.
- Line people are not necessarily responsible for poor quality.
- Management—and only management—can improve quality.

QUALITY CONTROL PROGRAMS

The Shewhart control chart technique was developed in 1924 and has been in continuous use since then. Perhaps the only fundamental change in the Shewhart chart has been the simplification evolved by mathematical statisticians by which control charts could be determined using the range of observations, rather than more time-consuming calculations for standard deviations for each subgroup. An evaluation of Shewhart's statistical approach was published in 1939 by W. E. Deming, who later defined "constant-cause systems, stability, and distribution" in simple terms to show how a control chart determined when a process was in statistical control. After over 50 years, these principles are still valid and are the basis for most of the successful quality control programs in use today.

One of the philosophies attributed to Dr. Deming is that judgment and the eyeball are almost always wrong. X-bar, $R,$ and p charts are the only evidence that a process is in control. Failure to use statistical methods to discover which type of cause (common or system cause versus special or assignable cause) is responsible for a production problem generally leads to chaos. Statistical methods, properly used, direct the efforts of all concerned toward productivity and quality.

Dr. Deming stated that 85% of the causes of quality problems are faults of the system (common causes) which will remain with the system until they are reduced by management. Only 15% of causes are special or assignable causes specific to an individual machine or worker and are readily detectable from statistical signals. Confusion between common and assignable causes leads to frustration at all levels and results in greater variability and higher costs—the exact opposite of what is needed. Without the use of statistical techniques, the natural reaction to an accident, a high reject rate, or a production stoppage is to blame the operators.

The worker is powerless to act on a common cause. The worker has no authority to sharpen the definition and tests determining what constitutes acceptable quality, cannot do much about test equipment or machines that are out of order, cannot change the policy of specifications for procurement of incoming materials, and is not responsible for design of the product.

Several quality control leaders have developed formalized multistep programs. It is difficult to look at a summary of these steps and determine which system is best for a given company, because the programs must be tailored to specific situations. Note how even these recognized authorities disagree on certain measures. A summary of the steps suggested by various quality control authorities follows. They are not complete descriptions, but they serve to differentiate the emphasis of each program.

J. M. JURAN (JURAN AND GRYNA, 1970)
1. Establish quality policies and guides for management action.
2. Establish quality goals.
3. Design quality plans to reach these goals.
4. Assign responsibility for implementing the plans.
5. Provide necessary resources.
6. Review progress against goals.
7. Evaluate manager performance against quality goals.

W. E. DEMING (DEMING, 1986)
1. Create constancy of purpose toward improvement of product and services.
2. Adopt the new philosophy that we are in a new economic age.
3. Cease dependence on mass inspection as a way to achieve quality.
4. End the practice of awarding business on the basis of price tag.
5. Constantly and forever improve the system of production and service; the system includes the people.
6. Institute on-the-job training.
7. Provide leadership to help people and machines do a better job.
8. Drive out fear.
9. Break down barriers between departments.
10. Eliminate slogans and targets for zero defects and new productivity levels.
11. Eliminate work standards and management by objectives.
12. Remove barriers that rob people of their right to pride of workmanship.
13. Institute a vigorous program of education and self-improvement.
14. Put everyone in the company to work to accomplish the transformation.

ARMAND V. FEIGENBAUM (FEIGENBAUM, 1983)
1. Agree on business decisions at the boardroom level to make quality leadership a strategic company goal and back it up with necessary budgets, systems, and actions. Each key manager must personally assess performance, carry out corrective measures, and systematically maintain improvements.
2. Create a systemic structure of quality management and technology. This makes quality leadership policies effective and integrates agreed-upon quality disciplines throughout the organization.
3. Establish the continuing quality habit. Today's programs seek continually improving quality levels.

TOM PETERS (PETERS AND WATERMAN, 1982)
1. Offer abiding management commitment.
2. Institute wholesale empowerment of people.
3. Involve all functions—and allies—of the firm.
4. Create encompassing systems.
5. Pay more attention to customer perceptions than to technical specifications.

P. B. CROSBY (CROSBY, 1979)
1. Management commitment.
2. Quality improvement team.

3. Quality measurement.
4. Cost of quality evaluation.
5. Quality awareness.
6. Corrective action.
7. Establish an ad hoc committee for the Zero-Defects Program.
8. Supervisor training.
9. Zero Defects Day.
10. Goal setting.
11. Error cause removal.
12. Recognition.
13. Quality councils.
14. Do it all over again.

M. R. HUBBARD (HUBBARD, 1987)
1. Select an area within the company in need of assistance, using Pareto or political procedures.
2. Using statistical techniques, calculate process capability and determine control limits.
3. Establish sampling locations, frequency, size, and methods of testing and reporting. Study process outliers and determine assignable causes of variation at each process step.
4. Correct assignable causes by modifying the process, and calculate improved process capability. Report to management dollars saved, improved productivity, reduced scrap, rework, spoilage, product giveaway, overtime, and so on.
5. Repeat Steps 3 and 4 until no further improvements are apparent.
6. Design experiments modifying the process to further improve productivity, and follow by returning to Step 2, using the most promising test results.
7. Move on to another area of the company (another line, another function, another department) until the entire company is actively pursuing quality programs.
8. Where possible, install quality attribute acceptance sampling plans as a safeguard for quality in the process and the finished product. Expand this into a company-wide audit system.

TOTAL QUALITY MANAGEMENT PRACTICES (U.S. GENERAL ACCOUNTING OFFICE, 1991).
1. Customer-defined quality.
2. Senior management quality leadership.
3. Employee involvement and empowerment.
4. Employee training in quality awareness and quality skills.
5. Open corporate culture.
6. Fact-based quality decision-making.
7. Partnership with suppliers.

HAZARD ANALYSIS CRITICAL CONTROL POINT (DEPT. HEALTH
AND HUMAN SERVICES—FDA. 1994)
1. Identify food safety hazards.
2. Identify critical control points where hazards are likely to occur.
3. Identify critical limits for each hazard.
4. Establish monitoring procedure.
5. Establish corrective actions.
6. Establish effective record-keeping procedures.
7. Establish verifying audit procedures.

ISO 9000 STANDARDS (INTERNATIONAL ORGANIZATION FOR
STANDARDIZATION REVISED 1994)
1. ISO 9000. Quality Management and Quality Assurance Standards—
   Guidelines for Selection and Use.
2. ISO 9001. Quality Systems—Model for Quality Assurance in Design/De-
   velopment, Production, Installation, and Servicing.
3. ISO 9002. Quality Systems—Model for Quality Assurance in Production
   and Installation.
4. ISO 9003. Quality Systems—Model for Quality Assurance in Final Inspection and Test.
5. ISO 9004. Quality Management and Quality System Elements—Guidelines.

MALCOLM BALDRIDGE NATIONAL QUALITY AWARD (U.S. DEPT. OF
COMMERCE 1987, AND REVISED ANNUALLY)
The seven categories on which these quality awards are based have been revised
over the years. Two years have been selected at random as examples.

1990 EXAMINATION CATEGORIES	1995 EXAMINATION CATEGORIES
1. Leadership	1. Leadership
2. Information and Analysis	2. Information and Analysis
3. Strategic Quality Planning	3. Strategic Planning
4. Human Resources Utilization	4. Human Resource Development and Management
5. Quality Assurance of Products and Services	5. Process Management
6. Quality Results	6. Business Results
7. Customer Satisfaction	7. Customer Focus and Satisfaction

Computer-Integrated Management (Approximately 1987)

Computer-integrated management (CIM) is a system designed to control all
phases of a food process by the use of computers. The goal is to utilize computer
power in product design, engineering, purchasing, raw material control, production scheduling, maintenance, manufacturing, quality control, inventory control,

warehousing and distribution, cost accounting, and finance. By integrating the data bases and commands of the individual computer systems throughout all of these stages, it should be possible to improve the efficiency of production planning and control, decrease costs of each operation, and improve both process control and product quality. The goal is to optimize the entire system through the use of computerized information sharing.

Other Quality Programs

Since the 1980s, several additional techniques have been offered with the goal of improving quality control programs. For the most part, they are business process tools rather than statistical quality control techniques, but have often had favorable effects on both productivity and product quality. Some of the many examples are Total Quality Management, Teams, Reengineering, Benchmarking, Empowerment, Continuous Improvement, Quality Function Deployment, Computer Applications, Computer Controlled Processes, Computer Analyses of Data, Computerized SPC, Real-Time Manufacturing, Expert Systems, and so forth.

PROBLEMS WITH TOOL SELECTION

The difficulty of selecting the correct statistical tools for problem solving might be illustrated by the analogy of selecting the correct tools for a painting project. Assume that we wish to paint a rod. We are faced with the following choices:

   5 finishes: flat, satin flat, satin, semi gloss, gloss
   3 solubilities: oils, water, organic solvents
   5 types: lacquers, enamels, stains, primers, fillers
   4 spreaders: air and pressure spray, roll, brush, gel brush

Although all of these possibilities are not necessarily combinable, there is a possibility of at least $5 \times 3 \times 5 \times 4 = 300$ combinations of kinds of paints and applicators.

This does not include all paint types or the myriad of shades available; nor does it consider the special formulations for floor, ceiling, deck, wall, furniture, concrete, metal, glass, anti-fouling, and so on. (In the metal category alone, there are subclasses such as iron, galvanized, copper, aluminum, and others.)

Knowing which tools are available may not solve our painting problem. Any paint will cover wood, concrete, or some metals, but will it peel, blister, fade, discolor, weather, corrode, flake, or stain? Will it leave brush marks or a slippery surface? Is it toxic? Does it have an unpleasant odor? Will it resist a second coat?

In short, knowing which tools are available is necessary. Knowing the proper use of these tools is absolutely imperative.

QUALITY CONTROL TOOLS

Following is a list of the most common statistical tools used in quality control applications. These will be covered in greater detail later.

STATISTICAL TOOL	USE
Acceptance Sampling Plans	Evaluate product attribute quality
Analysis of Variance	Establish significance of difference between two sets of data
Cause and Effect Chart	Display sources of problems
Confidence Limits	Determine reliability of sample results
Check Sheet	Locates major problems
Control Charts	Early detection of process variability
Cusum Chart	Cumulative subgroup difference plot
Design of Experiments	Provide valid data with minimum test
Evolutionary Operation	Shortcut response surface testing
Flow Chart	Defines process
Histogram	Process frequency distribution
Pareto Chart	Display frequency of problem areas
Probability Distributions	Summarize large groups of data
Process Capability	Level of yield uniformity
Regression Analysis	Determine mathematical relationships between two sets of variables
Sample Size Probabilities	Sample size determination
Scatter Diagram	Shows relationship of variables
Statistical Inference	Significance of data difference
Taguchi Method	Specification and tolerance technique
Trend Chart	Systematic drift analysis

These tools have specific applications in industry, and care should be taken to select the proper ones. Computers provide a convenient and speedy method of converting large volumes of data into charts and summaries with uncanny accuracy, but the use of the incorrect program or the selection of an incorrect tool in a program can lead to confusion. Table 1-1 outlines some of the most common applications of statistical techniques for quality control in the food industry.

Seven basic tools have been used successfully in food quality control programs for decades and, in all likelihood, will remain as the foundation for future quality needs in the industry. Over many years, there has been general agreement (*Quality Progress,* 1990) that these seven tools should be in every quality control program. They are discussed in some detail later. Following is a list with a brief explanation and a simplified example of each.

Table 1-1. Use of SQC Techniques in the Food Industry

Applications	Use technique
PRODUCT DESIGN Specifications: Product, ingredient, packaging Sensory evaluation Function, shelf life	A
PROCESS DESIGN Process specification, process capability	A, B
VENDOR Selection, qualification, control	B, C
INCOMING QUALITY Materials, supplies	B, C
PROCESS SPECIFICATION CONFORMANCE Sort, clarify, wash, heat, filter, cool, mill, other Package integrity, code, fill, appearance Microbiology	B, C, E, F
PRODUCT SPECIFICATION CONFORMANCE Sensory: Color, flavor, odor Structure, function	B, C
STORAGE AND DISTRIBUTION	B, C
CONSUMER COMPLAINTS	B, D
AUDIT Laboratory control Process, product, field performance	A, B, C
PRODUCT, PROCESS IMPROVEMENT	A, D, F

Note: Statistical techniques
A. Experimental design: Factorial, ANOVA, Regression, EVOP, Taguchi
B. Control charts: X-bar, R, p, np, c
C. Acceptance sampling: Attributes MIL STD 105E, Variables MIL STD 414
D. Diagrams: Pareto, Cause and Effect
E. Special sampling: Skiplot, Cusum, Scatter Diagram, Flow Chart, Histogram, Check Sheet
F. Special charts: Sequential, Trend Analysis, Consumer Complaints

1. Flow Chart: A picture of a process, using engineering symbols, pictures, or block diagrams, which indicates the main steps of a process. (See Fig. 1-1.)

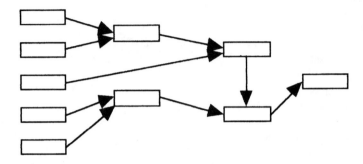

Figure 1-1. Flow chart.

2. Cause and Effect Diagram: A pictorial representation of the main inputs to a process, problem or goal, with detailed subfeatures attached to each of the main inputs. (also referred to as Ishikawa or fishbone diagrams.) (See Fig. 1-2.)

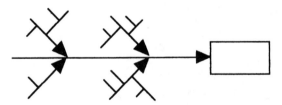

Figure 1-2. Cause and effect diagram.

3. Control Chart (variable and attribute): A graph of a process characteristic plotted in sequence, which includes the calculated process mean and statistical control limits. (See Fig. 1-3.)

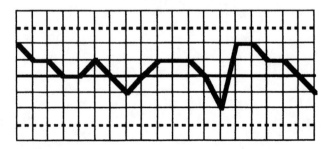

Figure 1-3. Control chart.

4. Histogram: A diagram of the frequency distribution of a set of data observed in a process. The data are not plotted in sequence, but are placed in the appropriate cells (or intervals) to construct a bar chart. (See Fig. 1-4.)

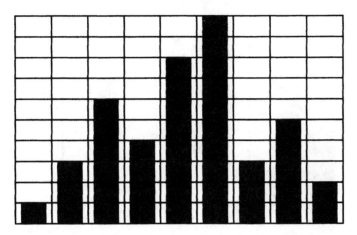

Figure 1-4.   Histogram.

5. Check Sheet: Generally in the form of a data sheet, used to display how often specific problems occur. (See Fig. 1-5.)

Figure 1-5.   Check sheet.

6. Pareto Charts: A bar chart illustrating causes of defects, arranged in decreasing order. Superimposed is a line chart indicating the cumulative percentages of these defects. (See Fig. 1-6.)

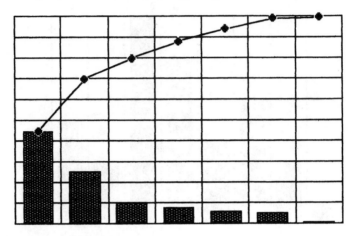

Figure 1-6.   Pareto chart.

7. Scatter Diagrams: A collection of sets of data which attempt to relate a potential cause (X-axis) with an effect (Y-axis). Data are collected in pairs at random. (See Fig. 1-7.)

Figure 1-7.   Scatter diagram.

2   Food Quality Systems

As a company grows, the need for formal departmental operating procedures and reports generally produces a large volume of standard manuals. The quality department in a food manufacturing company may be the last department to assemble such a written system, and there are perhaps more excuses than reasons for this:

- Products change from year to year, and someone would have to be retained on the quality department payroll just to keep up with the paperwork. (Unrealistic.)
- The food industry is regulated by federal agencies (Food and Drug, Commerce, Agriculture, and others) and by state and local agencies (Weights and Measures, Public Health, and others). Therefore, there is no need to further formalize quality procedures. (Untrue.)
- A food processing company could not remain in business unless its quality systems were adequate. It might be risky to change the existing system. (Head in the sand.)
- It is necessary to remain flexible in the food business so that the company can take advantage of new developments quickly. A formalized system tends to slow things down. (Absence of a system may bring things to a standstill.)

These excuses might be used by other departments within a food company as well (accounting, personnel, or distribution), but companies generally have rigorous formalized procedures for these departments. The quality department has frequently been overlooked in this respect, partly because it is a relatively new discipline. The *Food Processing Industry Quality System Guidelines* was prepared for the American Society for Quality Control in 1986. Until then, the Society (the professional organization dedicated to promotion of quality control in industry) had no recommendations specifically for the food industry.

Perhaps a more common reason for the lack of system emphasis on quality is that the techniques of statistical quality control are not well understood by upper management. Until the 1980s, few colleges or universities offered degrees in statistical quality control; few even offered classes on the subject. As a direct result, quality control was too often mistakenly thought to be concerned with inspection, sorting, sanitation management, and monitoring the retorting process for low-acid canned foods.

Quality control in the food industry was regarded as a cost center contributing to overhead, rather than a potential profit center contributing to savings. Following a series of successes at home and abroad in quality control—quality improvement, process improvement, productivity improvement, reduction in cost of scrap, rework, and product giveaway—the reputation of properly organized and operated quality control departments has gradually changed; what was once viewed as a cost generator is now seen as a profit center.

A quality system which starts at the product concept and development stage has the greatest opportunity to reduce new product costs. The cost of remedying design failures is minimized when shortcomings are detected at the concept or prototype stages of development. The cost of remedying failures rises exponentially at later stages (pilot plant, production run, or market distribution). A documented system can assure that geographically separated divisions of a company know how to produce uniform product quality and can communicate process improvement information between plants. Such a system provides an effective tool for training new employees both inside and outside the quality department. Perhaps most important, a documented quality system can be created to emphasize continuously the twin goals of attaining uniform quality and profit improvement.

SEVEN QUALITY SYSTEM GUIDELINES

Chapter 1 outlined nearly a dozen approaches to developing a quality system. A more detailed discussion of some of the more recent system guidelines follows. It should be emphasized that the seven basic tools of quality should be included in each of these systems.

Food Processing Industry Quality System Guidelines

The generic guidelines for quality systems developed by the American National Standards Institute (ANSI Z1.15) provide an excellent basis for establishing effective quality control systems, but they are geared more toward hardware manufacture than food processing. A committee of food quality experts, chaired by Sydney Pearce, restructured this standard for use by the food processing industry and published the guidelines in 1986. This document covers the following:

- Administration—includes quality policy, objectives, quality system, planning quality manual, responsibility, reporting, quality cost management, and quality system audits.

Each of these subjects is covered in detail. For example, *quality system* provides for individual policies, procedures, standards, instructions, and so on, covering the following: ingredients, packaging, processing, finished product, distribution, storage practices, vendor/contract processors relations, environmental standards, sanitation, housekeeping, pest management/control, shelf life, design assurance, recall, quality costs, user contacts, complaint handling and analysis, corrective action, motivational and training programs, and others.

- Design Assurance and Design Change Control—contains 12 subsections, such as concept definition, design review, and market readiness review.
- Control of Purchased Materials—an excellent summary of supplier certification requirements, such as specifications, system requirements, facility inspection, and assistance to suppliers.
- Production Quality Control—contains 24 detailed requirements under the following subheadings:
  Planning and controlling the process
  Finished product inspection
  Handling, storage, shipping
  Product and container marking
  Quality information
- User Contact and Field Performance—includes product objective, advertising, marketing, acceptance surveys, complaints, and analysis.
- Corrective Action—covers detection, documentation, incorporating change, product recall, and nonconforming disposition.
- Employee Relations—selection, motivation, and training.

Good Manufacturing Practice

Although not one of the *statistical* tools of quality control, Good Manufacturing Practices (GMPs) belong in every food quality control system. The Code of Federal Regulations (CFR 21 Part 110, GMP) provides excellent definitions and criteria which determine if product has been manufactured under conditions which make it unfit for food, or if the product has been processed under unsanitary conditions resulting in contamination with filth, or otherwise rendered injurious to health. It contains detailed requirements for avoiding these possibilities in the following general areas:

Personnel—disease control, cleanliness, education and training, and supervision.
Plant and Grounds—proper equipment storage, maintenance of surrounding property, effective systems for waste disposal, space for equipment placement and storage of materials, separation of operations likely to cause contamination, sanitation precautions for outside fermentation vessels, building construction to permit adequate cleaning, and adequate lighting, ventilation, and screening.

Sanitary Operations—building and fixtures: maintenance, cleaning and sanitizing to prevent contamination; special precautions for toxic sanitizing agents; pest control. Food contact surfaces: sanitation procedures.

Sanitary Facilities and Controls—water supply, plumbing, sewage disposal, toilet facilities, hand-washing facilities, and rubbish and offal disposal.

The Code then follows with specific GMP regulations for equipment and for process controls.

Equipment and Utensils—Design, materials and workmanship shall be cleanable, protected against contamination, and shall be nontoxic, seamless, and properly maintained. (Some specific types of equipment are referred to: holding, conveying, freezing, instrumentation, controls, and compressed gases.)

Processes and Controls—Adequate sanitation in receiving inspection, transporting, segregating, manufacturing, packaging, and storing. Appropriate quality control operations to ensure that food is suitable for human consumption and that packaging materials are safe and suitable. Assigned responsibility for sanitation. Chemical, microbial, and extraneous material testing. Reject adulterated or contaminated material.

- Raw materials: Shall be inspected for suitability for processing into food. Stored to minimize deterioration. Wash and conveying water to be of adequate sanitary quality. Containers shall be inspected for possible contamination or deterioration of food. Microorganism presence shall not be at a level which might produce food poisoning and shall be pasteurized during manufacturing to maintain a safe level. Levels of toxins (such as aflotoxin), or presence of pest contamination or extraneous material to be in compliance with FDA regulations, guidelines, or action levels. Storage of raw materials, ingredients, or rework shall be protected against contamination and held at temperature and humidity which will prevent adulteration. Frozen raw materials shall be thawed only as required prior to use and protected from adulteration.

- Manufacturing operations: Equipment, utensils, and finished food containers to be sanitized as necessary. Manufacturing, packaging, and storage to be controlled for minimum microorganism growth, or contamination. (A number of specific suggestions for physical factors to be controlled: time, temperature, humidity, water activity, pH, pressure, flow rate. Controls for manufacturing operations are also suggested: freezing, dehydration, heat processing, acidification, and refrigeration.) Growth of undesirable organisms shall be prevented by refrigeration, freezing, pH, sterilizing, irradiating, water activity control. Construction and use of equipment used to hold, convey or store raw materials, ingredients, work in process, rework, food, or refuse shall protect against contamination. Protection against inclusion of metals or other extraneous material shall be effective. Adulterated food, ingredients, or raw materials shall be segregated and, if reconditioned, shall be proven to be effectively free from adulteration.

Mechanical manufacturing steps such as washing, peeling, trimming, cutting, sorting, inspecting, cooling, shredding, extruding, drying, whipping, defatting, and forming shall be performed without contamination. Instructions are offered for blanching, with particular emphasis on thermophilic bacteria control.

Preparation of batters, breading, sauces, gravies, dressings, and similar preparations shall be prepared without contamination by effective means such as contaminant-free ingredients, adequate heat processes, use of time, and temperature controls, physical protection of components from contaminants which might be drawn into them during cooling, filling, assembling, and packaging.

Compliance may be accomplished by a quality control operation in which critical control points are identified and controlled during operation; all food contact surfaces are cleaned and sanitized; all materials used are safe and suitable; physical protection from contamination, particularly airborne, is provided; and sanitary handling procedures are used.

Similar requirements are listed for dry mixes, nuts, intermediate moisture food, dehydrated foods, acidified foods, and ice-added foods. Finally, the regulation forbids manufacturing human and nonhuman food grade animal feed (or inedible products) in the same areas, unless there is no reasonable possibility for contamination.

- Warehousing and Distribution: Storage and transportation of finished foods shall be protected against physical, chemical, and microbial contamination, as well as deterioration of the food and the container.

Food processing companies with sufficient staff might consider incorporating the GMP regulation in the quality control manual and conducting routine audits to assure conformance. Consulting firms are available to perform periodic GMP inspections for smaller organizations. In either case, a file of satisfactory audits could prove invaluable in the event of suspected product failure resulting in litigation.

It should be noted that the FDA regulations for Good Manufacturing Practice is modified from time to time, and it is necessary to periodically review quality control procedures to ensure compliance.

ISO 9000 Standards (International Organization for Standardization revised 1994)

The International Organization for Standardization in Geneva, Switzerland began to develop a series of standards to describe an ideal, generic quality system in the late 1970s. Based on the British quality standard, the initial intent was to clarify contracts between suppliers and their customers. It became apparent after a few years that companies which were registered for compliance with these standards would not require most supplier audits, as customers could be assured of product which would meet their specifications. Some countries have expanded this concept to require that imported goods must be produced under ISO standards. Although the ISO requires that all standards be reviewed and updated every 5 years, it is expected that changes will be modest and gradual. The 1994 changes, for example, include relatively minor format and wording, a greater emphasis on documented procedures (including the quality manual), a few small additions to management responsibility and staffing, addition of a quality planning document, and a few definitions such as verification and validation.

The 20 elements of ISO 9001 are a blend of many of the basic features of the GMP (Good Manufacturing Practices) and the HACCP (Hazard Analysis Critical Control Point) regulations of the FDA. In the introductory paragraphs of the standards it is stated that they are generic and are not intended to enforce uniformity of quality systems, because the needs of an organization, its objectives, the products and services provided, and the processes and specific practices used may be unique.

Of the five ISO standards listed in Chapter 1, ISO 9001 is perhaps the most comprehensive, and provides guidance for 20 elements of a quality system applicable to the food industry. ISO 9004 is a more generalized quality management document. The other standards are perhaps more suited to hardware manufacturing industries, in that there is some emphasis on such areas as design review, reliability, configuration management, failure modes, product maintainability and service, none of which are particularly applicable to food products.

Following are outlines of the salient features of ISO 9001 and ISO 9004 which apply to the food industry.

ISO 9001. The 9001 standard requires the establishment and documentation of a quality system as a means of ensuring that the product meets specified requirements. It is titled *Quality Systems—Model for quality assurance in design, development, production, installation and servicing.* It states that a quality manual shall be prepared in conformance with the standard, be consistent with the company's stated quality policy, and shall be effectively implemented.

1. Management responsibility
   Define quality policy and quality objectives.
   Document responsibility for specific quality functions.
   Identify resources for performance, verification, and audits.
   Appoint management representative for system review.
2. Quality system
   Prepare quality manual covering ISO 9001 requirements.
   Document quality plans and procedures.
   Identify controls, processes, equipment, and skills.
   Update controls and instrumentation.

Clarify standards and specifications.
Describe quality records required.
3. Contract review
Review each specific order (from a customer) to ensure agreement on requirements and specifications.
4. Design control
(The terminology in ISO 9001 describing requirements of design control is more applicable to the hardware industry than to the food processing industry. The terms below have been changed to those commonly used in food research and development activities.)
Prepare written plans for each research activity.
Define organizational and technical interfaces.
Document statutory and regulatory requirements.
Prescribe acceptance criteria.
Identify critical characteristics (safety, storage, etc.).
Document research reviews during development.
Verify that research at each stage meets requirements.
Validate new product or process on production equipment.
5. Document and data control
Maintain all documents referred to in ISO 9001.
Prepare and update master list.
Invalidate and remove obsolete documents.
Identify all document changes.
Perform changes with original review personnel.
6. Purchasing
Document purchasing procedures.
Include specifications on purchasing documents.
Specify quality standard and system to be applied.
Select subcontractors on ability to meet quality specs.
Define control over and verification of subcontractor quality.
Maintain records of subcontractor quality tests and audits.
7. Control of customer-supplied product
Document procedures for purchased supplies handling.
   Verify, store, maintain.
8. Product identification and traceability
Establish procedure to identify batches throughout process.
9. Process control
Document procedures which affect quality.
   Use suitable production equipment and environment.
   Ensure compliance with regulations and standards.
   Monitor suitable process parameters.
   Monitor suitable product characteristics.
   Approve processes and equipment.
   Standards for workmanship.

Suitable equipment maintenance.

Maintain process records.

10. Inspection and testing

    Document and maintain procedures for inspection and test.

    Quarantine incoming material until quality is verified.

    Inspect and test in-process material as required by the plan.

    Inspect finished product as required by quality plan.

    Hold finished product until authorized for release per plan.

    Maintain inspection and test records.

11. Control of inspection, measuring, and test equipment

    Document procedures to control, maintain, and calibrate inspection, measuring, and test equipment and software used to provide data confirming product conformance to specifications.

12. Inspection and test status

    Identify status of product inspection and test procedures throughout production to ensure that only product which conforms to specifications has been shipped.

13. Control of nonconforming product

    Document procedures to isolate nonconforming product.

    Identify authority for disposition: rework, regrade, and scrap.

    Reinspect reworked or regraded product.

14. Corrective and preventive action

    Document procedures for corrective and preventive action.

    Record procedure changes resulting from these actions.

    Corrective action includes:

       Effective customer complaint handling.

       Investigation of causes for nonconformities.

       Determination of corrective action.

       Application of controls to assure effectiveness.

    Preventive action includes:

       Review of procedures and records.

       Determination of steps to deal with prevention.

       Initiation of preventive action and use of controls.

15. Handling, storage, packaging, preservation, and delivery

    Document procedures.

    Maintain records.

16. Control of quality records

    Document procedures for identification, collection, indexing, access, filing, storage, maintenance, and disposition of quality records.

    Establish and record retention times for each class.

    Records may be hard copy or electronic.

17. Internal quality audits

    Document procedures for conducting internal quality audits.

    Schedule on the basis of importance.

    Select auditors independent of the area involved.

Report audit results to personnel in area audited.

Management personnel shall take corrective action.

Follow-up procedures shall verify effectiveness of action.

18. Training

Document training needs for personnel affecting quality.

Provide such training and maintain records.

Assign specific tasks to those qualified by training.

19. Servicing

Probably not applicable in food processing, except in the raw material, restaurant, and food service industry.

20. Statistical techniques

Ensure the proper use of statistical techniques where applicable in establishing, controlling, and verifying process capability and product characteristics.

ISO 9004.   International Standard ISO 9004 deals with quality management and detailed quality system elements. This is in contrast to ISO 9001, which is intended to demonstrate to its customers the quality control capability of a supplier. This standard is titled: *Quality Management and Quality System Elements—Guidelines.*

1. Management responsibility

Develop and state its corporate quality policy.

Define quality objectives at various levels of management.

Establish and implement a quality system.

2. Quality system principles

Include all functions and departments of the organization.

Define customers' expectations and product requirements.

Provide optimum cost specifications.

Management is ultimately responsible for all quality aspects.

Responsibility, authority, and communications defined.

Identify problem and solutions—actual and potential.

Provide resources: human, equipment, and software.

Establish competence and training requirements.

Organize system to provide continuous control.

Emphasize preventive actions.

Coordinate quality goals of various departments.

Document quality policies and procedures.

Issue quality manuals for corporation, divisions, project, work, and so on.

Prepare quality plans for new activities.

Maintain quality records (see below).

Provide quality audit plan: personnel, activities, and procedures.

Provide audit procedures, report, and follow-up.

Provide for independent quality system reviews.

3. Quality economics

Measure and report prevention, appraisal, and failure costs.

Define external quality costs.

4. Quality in marketing
   Marketing provides product need, grade, price, and timing.
   Define requirements: usage, sensory, standards, and packaging.
   Determine need for quality verification.
   Provide customer feedback data continuously.
5. Quality in specification and design
   Translate customer needs into specifications and processes.
   Assign design responsibilities.
   Determine schedule and checkpoints.
   Establish quality characteristics and methods of test.
   Document tests and evaluations of prototypes.
6. Design review
   This section is more applicable to hardware industries.
7. Quality in procurement
   Specify materials precisely.
   Select qualified suppliers by samples, plant visit, and so on.
   Agree to method for quality assurance, verification, and reject.
   Maintain supplier and receiving quality records.
8. Quality in production
   Provide controls for materials, equipment, processes, procedures, software, personnel, supplies, utilities, environments.
   Document production operations.
   Conduct process capability studies.
   Provide standards: written, photographs, and samples.
   Establish quality verification stages.
   Document test and inspection procedures.
   Encourage quality and process improvements.
9. Control of production
   This section is more applicable to hardware industries.
10. Product verification
    This section is more applicable to hardware industries.
11. Control of measuring and test equipment
    This refers to establishing procedures to monitor gauges, instruments, sensors, computer software, process instrumentation, and operator skills. (The food industry would also include laboratory ovens, incubators, standards, test chemicals, and bacteriological materials.)
    Specify corrective action.
    Designate outside testing facilities.
12. Nonconformity
    Identify nonconforming items or lots.
    Segregate and identify nonconforming items.
    Review for disposition: rework, reclassify, and scrap.
    Dispose of items as soon as practicable.
    Prepare and retain appropriate documents.

13. Corrective action
    Assign responsibility and authority.
    Determine impact on costs, performance, safety, and customer.
    Investigate possible causes of problems.
    Use statistical analysis.
    Change procedures and controls to prevent reoccurrence.
    Consider safety, liability, and customer satisfaction.
    Dispose by scrap, recall, and repair.
    Include permanent changes in procedure documents.
14. Handling and postproduction functions
    Handle and store to prevent damage.
    Identify with proper, legible, and durable markings.
    Identify required packaging in written instructions (for food industry: rotate stock).
    Establish early-warning system for product shortcomings.
15. Quality documentation and records
    File legible, dated documents as hard copy or in computer.
    Systematize removal of out-of-date documents.
    Control drawings, specifications, blueprints, inspection, instructions, test procedures, work instructions, operation, sheets, quality manual, operational procedures, and quality assurance procedures.
    Maintain quality records: inspection reports, test data, qualification reports, validation reports, audit reports, material review reports, and quality cost reports.
16. Personnel
    Provide training to personnel at all levels.
    Train executive management in quality tools and techniques, and criteria to evaluate the quality system.
    Provide quality training to personnel in quality, marketing, procurement, and process and product engineering.
    Emphasize statistical technique training: process capability studies, statistical sampling, data collection and analysis, problem identification, problem analysis, and corrective action.
    Train production workers in operation of instruments, tools and machinery, understanding documents provided, relationship of duties to quality and safety, and statistical analysis where needed.
    Arrange to qualify specialized personnel.
    Motivate personnel by explaining their tasks.
    Show effects of proper and poor performance.
    Apply motivation efforts at all levels of employees.
    Emphasize quality through an awareness program.
    Publicize satisfactory quality.
17. Product safety and liability
    Identify safety standards.
    (Other items are more applicable to hardware industries).

18. Use of statistical methods

Apply statistical methods in market analysis, product design, shelf life, process control, process capability studies, determination of quality levels and inspection plans, data analysis, performance assessment, and defect analysis.

Include design of experiments, factorial analysis, analysis of variance, regression analysis, risk analysis, tests of significance, quality control charts, cusum techniques, and statistical sampling inspection.

19. Appendices

One appendix covers a series of brief definitions of sampling inspection techniques, and basic statistical methods. A second appendix outlines product liability and user safety.

Malcolm Baldridge National Quality Award

The Malcolm Baldridge National Quality Award is presented by the President of the United States to a limited number of winners in manufacturing, service, and small companies each year. The primary purpose of the award is to recognize U.S. companies for business excellence and quality achievement. Additionally, the award has encouraged thousands of organizations to undertake quality improvement programs. As mentioned in the previous chapter, the Malcolm Baldridge National Quality Award categories have undergone significant changes since inception. The early guidelines focused on the applicant's *total quality system*. After a few years, this was modified to recognize *business excellence and quality achievement.* Either approach provides a useful format for evaluating and improving a company's quality system. However, it is suggested that companies use the 1990 Baldridge Award Application Guidelines as a model if they are not yet satisfied that they have achieved a detailed and effective quality system. Category 5 of this early guideline, titled Quality Assurance of Products and Services, covers seven of the basic requirements:

1. Design and introduction of quality products and services, emphasizing how processes are designed to meet or exceed customer requirements.
2. Process and quality controls which assure that the products and systems meet design specifications.
3. Continuous improvement techniques (controlled experiments, evaluation of new technology, benchmarking) and methods of integrating them.
4. Quality assessments of process, products, services and practices.
5. Documentation.
6. Quality assurance, quality assessment, and quality improvement of support services and business processes (accounting, sales, purchasing, personnel, etc.).
7. Quality assurance, quality assessment, and quality improvement of suppliers (audits, inspections, certification, testing, partnerships, training, incentives, and recognition).

Once a company is satisfied that an effective quality system has been established, it might be desirable to examine the Baldridge Award criteria of subsequent years to see if the goals of business excellence and quality achievement are also being met.

Total Quality Management

Total Quality Management (TQM) is more of a quality philosophy than a quality system. TQM is a goal in which every individual in every department of an organization is dedicated to quality control and/or quality improvement. In addition, this web of management may encompass suppliers and distributors as well. Some plans include the customers as partners in the quality efforts. Each organization has the freedom to define "total," "quality," and "management" in any way they believe will enhance the quality efforts within their scope.

One company instituted a program to train all of their employees with team techniques (see below) for solving quality problems and improving product quality. They referred to this activity as their TQM process.

Some companies have tried to avoid the stigma of reports of failed TQM programs by using other terminology: "The Universal Way," "Quality Improvement Process," "Total Quality Service," " Quality Strategies," "Leadership Through Quality," "Total Systems Approach," "Quality Proud Program," "Customer Driven Excellence," "Total Quality Commitment," "Total Quality Systems," and "Statistically Aided Management."

Food Technology magazine, published by the Institute of Food Technologists, had several articles with widely varied definitions of TQM. One all-encompassing description of TQM states that it is a way of managing business and is based on the American Supplier Institute approach:

- Incorporate leadership, cooperation, partnerships, and trust.
- Aim for long-term goals of the business.
- Concentrate on improving processes to improve results.
- Deploy means and goals to all levels of the organization.
- Use statistical quality control tools, quality function deployment, Taguchi methods, and other quality tools.

Quality Progress magazine, a publication of the American Society for Quality Control, devoted an entire issue (July 1995) to 13 articles which described and evaluated TQM. Each paper was required to submit a brief definition of TQM. Table 2-1 indicates the differences of opinion among the authors.

The observation that there is no consensus for a TQM definition should not lessen its virtues. Applying as many quality control principles in as many areas as possible is a desirable goal for any organization. Some companies have built successful TQM programs (using their own definition) by proceeding stepwise; others have attempted to formulate the entire concept at once, and implement it in stages. There probably is no one best way.

Table 2-1.   Differences in Definition of TQM

TQM system components	Author's definitions												
	1	2	3	4	5	6	7	8	9	10	11	12	13
Business or management philosophy	X					X							
Continuous (mind-set for) improvement					X	X	X	X		X		X	X
Continuous redesigned work process							X						
Customer focus/driven	X					X	X		X	X		X	X
Customer/supplier partnership				X									
Delight customers				X									
Develop improvement measure											X		
Develop with time				X									
Employee cooperation								X	X				
Evaluate continuous improvement		X	X										
Failure mode effects analysis					X								
Goal alignment					X								
Guarantee survival	X												
Improve customer satisfaction			X			X							
Improve financial performance	X												
Improve management system	X												
Improve materials and services supplied												X	
Learning architecture							X						
Meet customers' needs	X												
Meet employees' needs	X												
Meet owners' needs	X												
Mission driven									X				
Plan/do/check/act											X		
Reduce costs				X									
Reengineering				X									
Responsible individuals			X										
Responsive customer service									X				

Table 2-1.   (*continued*)

TQM system components	Author's definitions												
	1	2	3	4	5	6	7	8	9	10	11	12	13
Self-directed teams				X									
Set targets for improvement												X	
Statistical process control					X					X			
Strategic architecture		X											
Teamwork and trust			X				X						
Use of quality tools										X			

Team Quality Systems

Teams generally represent a subsystem of quality control and improvement systems. They rarely work well without a centralized quality control structure because they are usually task oriented. Team concepts and operation will be covered in some detail in a subsequent chapter.

Computer Network Quality Systems

Where operational cost/benefit analyses can justify the installation of computer systems on processing lines, opportunities for improved quality control accompany production gains. In addition to creating a system which can graphically show current progress of every step of a process, track the quantities of ingredients, supplies, work in process, and output, these same data signals can be harnessed to operate and control valves, burners, feeders, pumps, mixers, and other equipment. Additionally, quality control can have the ability to use real-time data for instantaneous control, to collect larger quantities of more meaningful and accurate data, to provide high-speed calculation and display of control charts, quality costs, and other quality reports, and to instantly create and access data storage. Computer systems are discussed in greater detail in Chapter 15.

Summary

The many quality system approaches outlined in Chapter 1 are still used in the food industry. There will never be complete agreement on the *one* best way, but recent trends have emphasized the use of portions of the older philosophies, along with several new ones. Those considered most effective at present include the eight systems explained above.

POLICY

A quality system is all but useless without an understanding of the company quality policy. If the company policy on product packaging is to provide inexpensive and functional sealed plastic film envelopes, then the quality control manager might be doing the company a costly disservice to insist on precise printing and color control of these envelopes. A verbal understanding of policy is not enough. The policy should be in writing, signed by a company official, and reproduced as the front page of the quality manual.

If no policy statement exists, it may be difficult to draw one up because of organizational peculiarities or sensitive personal relationships between department managers. One of the most effective methods of generating a policy document is for the quality manager to prepare a clearly labeled draft policy and forward it to the company president for revision and approval through whatever channels are considered necessary. This provides the president with the necessary guidelines, but still leaves him with complete authority to select the details. A simple policy statement might read:

> The Midwest Food Company is dedicated to supplying our customers with a competitive quality line of standard products and a superior line of choice quality products under the Gourmet label. All of our products shall be shelf-stable for one year, safe, nutritious, and flavorful, and shall compare favorably with the quality level of similar products found in our marketing area. They shall conform in all respects to the standards and regulations of federal, state, and local governments. The responsibility for quality control shall rest jointly with the managers of purchasing, production, distribution, and quality control.

Major specifics might be included:

> The quality control department manager shall establish and maintain product and process specifications. He or she shall be responsible for preparing and distributing routine quality reports, exception reports, quality improvement reports, and cost reduction proposals. He or she shall perform monthly audits to evaluate all phases of the quality control system, including sanitation, product performance, housekeeping, and quality training. Quality control meetings shall be chaired weekly by the chief executive officer, and attended by managers and lead foremen of all departments.

For some companies, more specific quality statements might be desirable, but generally, the longer and more complicated a policy statement, the less likely it will be read and followed. Details are best left for inclusion in the quality manual. Where there is potential for misunderstanding, it might be necessary for the policy statement to include the functional quality responsibilities of each department. The drawbacks to details of this nature are related to possible personality conflicts, turf battles, and restriction of the contributions of capable employees.

QUALITY CONTROL MANUAL

If a quality manual were prepared to conform to the best recommendations of all of the respected quality professionals, it would be so voluminous as to be useless.

It is essential, however, that all of the critical elements of a manual be prepared, updated, and—most important—used. One suggestion for minimizing size without compromising content is to divide the manual into a set of numbered submanuals:

MANUAL NUMBER	SUBJECT
Q1	Policy, Organization, Interrelationships
Q2	Research and Pilot Plant Quality Control
Q3	Material Quality Control
Q4	Production Quality Control
Q5	Product Performance
Q6	Corrective Action
Q7	Microbiological Procedures
Q8	Packaging Material Test Methods
Q9	Product Specifications
Q10	Quality Audit Procedures
Q11	Reports and Forms
Q12	Government Regulations
Q13	Product Recall System
Q14	Special Quality Requirements

An excellent guide to content and wording for an overall quality manual is Military Standard 9858A, obtainable from the Government Printing Office. Although it is hardware-industry oriented, it includes most of the basics and many details that might be overlooked.

The Secretary of Defense canceled MIL Q 9858A as of October 1996 and adopted the ISO 9000 system in its place. The basic principles offered in MIL Q 9858A are similar in most respects to the ISO 9000 standards, differing chiefly in the following areas:

a. MIL Q 9858A has generally been the responsibility of the Quality Control Department, whereas ISO 9000 cuts across organizational lines, requiring input and responsibility from all departments.
b. ISO 9000 quality systems require constant review and modification. In fact, the standard itself has a scheduled review and modification requirement.
c. ISO 9000 extends beyond the production quality concept and includes product design. It also covers, as previously noted, a number of activities which are more common to hardware manufacturers than food processors.

Test methods other than those specifically listed above might be included under other headings or compiled into a separate manual. Such tests are primarily concerned with raw materials, process, and product. In addition, sample size, selection, and frequency might be chosen from the above classifications, particularly if the techniques are complex. By preparing separate manuals to cover such subjects as sampling and testing, training of new employees is often simplified.

PRODUCT RESEARCH AND DEVELOPMENT

From a legal standpoint, one of the most important requirements in developing a new or improved product is documentation. This protects the company in the event of patent or copyright litigation. For the scientist or engineer, accurately prepared records provide a route map for others to follow and often suggest other developments for products, processing, or packaging to be explored at a later date.

Many creative food technologists prefer to work in an unstructured and informal atmosphere, and for small companies or short-term projects, formalized procedures are not required. Guidelines are intended to assist workers, not to bind them into a mold which restricts the inquisitive mind. Perhaps the systematizing of research and development (R&D) should be formalized more as an outline or checklist than a rigorous procedure. This permits flexibility and at the same time provides the safeguards which reduce the possibility of overlooking important development steps.

Quality Goal of R&D

The quality system should assure that specifications evolve which conform to marketing needs, manufacturing practices, and regulatory requirements, and cover all aspects of materials, process, packaging, storage, and use. These specifications might include the following:

Physical and sensory properties
Product function and nutritive values
Process equipment and process rates
Packaging and packaging equipment
Composition
Microbiological limitations
Shelf life
Labeling and coding
Regulatory requirements

Methods and Standards

A checklist for methods and standards might best be considered a flexible guide, because most development projects have special requirements and will not fit a rigorous mold. To avoid the possibility of overlooking important requirements, such a list should be reviewed at the start of any new development project:

Sampling and statistical analysis
Analytical testing methods
Calibration and maintenance of analytical equipment
Shipping tests and evaluation
Tolerance evaluation and review

Hazard and failure mode effect analysis
Specifications for material, process, product, package
Consumer test design and evaluation
Patent potential evaluation
Outline of quality system for initial production
FDA, USDA, EPA, and local regulations

Documentation

Progress reports are usually considered the basic documents for development studies, and their presentation should be well planned. Haphazard reporting may impede efforts by requiring disruptive staff meetings to explain reasons for failures or the direction of current research. At the very least, milestone reports should be made (where applicable) at six stages:

Product concept
Prototype development
Pilot plant trial
Qualification testing
Field evaluation
Market readiness

Reviews

Reports at the six stages above are intended to generate critical reviews by involved personnel. Criteria for acceptance of progress should be developed into a system, thereby permitting work to proceed to the next step. This should avoid unnecessary backtracking and interruption of the research process. A universal checklist is impractical because each research project tends to be unique, but this fact should not be used as an excuse to avoid detailed documentation. Perhaps a master list for qualification testing review might illustrate the scope of such a report:

Critical material
Production function and composition
Product stability: nutrient, sensory, function
Packing performance
Process variability
Process response: drift, interactions, errors
Shipping
Cost analysis

MATERIAL CONTROL

Just-in-time manufacturing systems set goals to lower costs by reducing inventories and eliminating rejects, rework, and downtime through acceptance of only defect-free raw materials in quantities no larger than those sufficient to keep the plant operating. Obviously, it is costly to produce satisfactory product from defective raw materials by sorting at the receiving dock or correcting on the line, and much effort has been made to assist vendors in producing and delivering defect-free field crops, food additives, and packaging materials. The key word here is *assist*. The farmer may be unaware of how critical the sugar–acid ratio of his fruit is in a jellymaking process. The packaging film supplier may be unaware of the relationship between film thickness uniformity and shelf life of a product. A fraction of a percent of gravel in unprocessed grain may abrade raw material cleaning equipment. By working with suppliers, a manufacturer can often reduce the quality costs of both parties. Improved materials reduce downtime, permit reduction of just-in-case inventories, and increase profits.

Material control starts, at least theoretically, at the product development stage. It becomes a real consideration during supplier (also referred to as "vendor") selection and eventually evolves into a daily consideration. Selection may be based on marginal criteria such as reputation or past performance. Where material specifications are clearly written, they may serve to eliminate suppliers who are unable either to reach or maintain the requirements. This should emphasize the need to be realistic when preparing specifications. Material specifications need not be stricter than is actually necessary in the process; this might rule out alternative sources and may often result in unnecessary added material cost to achieve an unrealistic level of quality.

By working together, people responsible for quality control, purchasing, and production can work with their counterparts in suppliers' organizations to prepare realistic purchase specifications and contractual requirements. With a clear understanding of the rules, suppliers might be willing to provide certification documents with each shipment, thus avoiding the need for major incoming inspection facilities at the processing plant. Under such conditions, suppliers might also be willing to accept financial responsibility for production losses due to nonconforming raw material resulting in line stoppages, rejected or spoiled product, lost packaging materials, repacking, and overtime costs. This has long been a practice for large-volume food processors and should apply equally well to smaller companies.

An assessment of the quality of materials from suppliers should include:

- Inspection of supplier quality control programs.
- Evaluation of supplier specifications.
- Examination of supplier facilities and production procedures.
- Evaluation of performance through periodic audits.

Although one goal of supplier–purchaser relationships is the elimination of incoming material inspection, it is unlikely that 100% achievement can be accomplished without serious risk. For both attribute and variable inspection plans commonly used, switching plans are available which permit reduced inspection when a supplier has provided satisfactory material over a specified number of lots.

Using these can substantially reduce material inspection costs while assuring that only usable material will be accepted. Ultimately, this might lead to a just-in-time supplier program.

If nonconforming incoming material is discovered, quality control, purchasing, and manufacturing departments should be informed immediately to avoid possible losses from production downtime or spoilage. Unless properly controlled, rejected materials can find their way onto the production floor. The simplest control system includes four distinct steps:

1. Document the identification of the shipment, inspection results, disposition recommended, and (when completed) final disposition.
2. Label nonconforming material clearly with distinctive markings and easy-to-read warnings that it should not be used.
3. Segregate the unsatisfactory shipment to prevent the possibility of mixing it with satisfactory product. If a shipment is contaminated, isolate it from areas where edible material is stored.
4. Dispose of promptly. If destruction of the material is deemed necessary, such destruction should be observed by quality control personnel.

Even under ideal conditions, minimal receiving inspection is always required for compliance with current FDA Good Manufacturing Practices regulations, other government regulations, and company policies. Likewise, supermarket chain stores and large food product distributors consider food processors their suppliers and are likely to require specific receiving inspection procedures at the processing plant as part of their overall quality systems. Regulatory or contractual considerations notwithstanding, it is imperative to inspect at the receiving dock for evidence of potential problems with sanitation, housekeeping, pest control, adulteration, or fitness-for-use. This represents a major departure from the practices of hardware manufacturing industries: the end product of a food processor is eaten!

PRODUCTION QUALITY CONTROL

The major efforts of quality control are devoted to the production line and must be designed around process capability studies (which is discussed in detail in a later chapter). A process capability study determines whether a process can meet desired product specifications, where sampling points should be established, and frequency of sampling cycles. In addition, the direction process change should take to improve quality and productivity is often highlighted during process capability studies. The need to periodically conduct confirming studies is often overlooked. When a process is running comfortably in control for a period of time, a repeat process capability study may indicate the need for new control limits—and perhaps new product specifications and new process targets. Normally, this is the basis for on-line product quality improvement.

To continuously control a process without written procedures is next to impossible. Word-of-mouth instructions from supervisors to operators are invariably modified, misunderstood, or forgotten. Control cannot be accomplished when a process operator explains his unusual procedures with excuses such as, "It works better this way," "I thought that's what I was supposed to do," "It didn't seem that important," or, worst of all, "Nobody ever explained that to me." Documented procedures are difficult to establish and require continuous upgrading and correction, but they are essential to providing a system which can be controlled to the same standards day after day. A second major benefit of written procedures is their availability for providing uniform training to all new employees.

No one person should be totally responsible for writing production operation procedures. Procedures are less likely to be followed if they are forced on the production department by an outsider, even if they have been prepared under the direction of top management. To ease the introduction and acceptance of procedures, all affected supervisors and managers should have the opportunity to review drafts as they are developed and be responsible for offering modifications to a draft before the final document is approved and used.

Occasionally, management might balk at the concept of providing written procedures of proprietary operations to employees who could allow them to fall into the hands of competitors. These fears might be alleviated by the use of codes for critical temperatures, pressures, flows, or special equipment. Where possible, such critical data could be left blank on the procedures but posted on the equipment for use by the employee. It is not unusual for employees with proprietary process knowledge to leave one company for employment with a competitor, but the advantages of written process procedures far outweigh the disadvantages.

Control of operations may include several functions, such as:

- In-process inspection of materials, product, and package.
- Process verification.
- Status identification of raw materials, partially processed product, and finished product (ordinarily by using a tag system) (see Fig. 2-1).
- Calibration of test equipment including meters, gauges, media, reagents, metal detectors, check-weighers, production instrumentation, and so on.
- Nonconforming material identification, isolation, and release. (An accurate history of nonconforming material must be maintained for the life of the product. Corrective action recommendations should be made a part of this system.)
- Document control, including procedures, specifications, test methods, formulations, and quality reports.
- Material identification, rotation, and protection from infestation or physical damage.

Some emphasis on test equipment calibration is in order. There is a natural tendency to program a computer, install a micrometer, or calibrate a scale, and then walk away believing that the job is finished. It is not. Others might change the cal-

STANDARDIZATION RECORD

| REAGENT_____ |
| DATE REC'D_____ |

TEST DUE	TEST RUN	OK	BY

HOLD TAG

LOT NO:_____
SUPPLIER:_____
DATE REC'D_____
REASON REJECTED_____

LOCATION:_____
DATE RETESTED_____
DISPOSITION_____
_____
_____
SIGNED_____ QC DEPT

Figure 2-1. Suggested forms for controlling reagent standardization and for controlling nonconforming materials.

ibration, vibration or wear might induce drift, or corrosion might affect readings. The reliability of the test equipment is as important as the quality control system itself. The simplest calibration system should encompass the following:

- Verification of accuracy before permitting use. All new measuring and testing equipment, chemicals, reagents, scale weights, microbiological media, and secondary standards should be verified or calibrated against certified standards, preferably having a known relationship to national standards.
- Periodic verification to be performed at defined intervals.
- Control over the verifications and standardizations. This shall include tagging of equipment (where practical) with the date tested and the next inspection date shown on the tag (see Fig. 2-1). Additionally, a complete and up-to-date record of the status of each measuring and testing item shall be maintained. The analytical methods used for standardization should be subjected periodically either to collaborative tests or to internal monitoring by control charts.

The final step in production quality control is finished-product inspection. Prior to the common use of statistical quality control, this single step was often considered a quality control system. Experience (and common sense) have shown that finished-product inspection is generally an audit of all quality subsystems used in creating the product. At this stage, the quality has already been built into the product, and nothing can be done other than accept, rework, or scrap. Even so, finished-product inspection is an important step that should not be overlooked when constructing a quality system.

It should be emphasized again that finished-product inspection alone is not quality control. It is an inefficient and ineffective method of assuring consistent quality. The individual responsible for finished-product testing can easily succumb to boredom and fail to see defects. Previously undiscovered flaws may be present but not observed because the inspector has not been conditioned to look for them. But final inspection cannot be ignored; it is the last chance to discover quality problems before the consumer discovers them.

Finished-product testing can include examination for whatever attributes are included in the specification, but the crucial test should be performance of the product. The product should look, smell, dissolve, and taste right. It should be tested in the same manner as the customer or consumer would normally use it. If it is a tea bag, the paper should not disintegrate in the cup—the customer is not interested in the number of grams of force required to separate the seam, but only that the tea leaves remain inside the bag.

Another class of finished-product inspection is acceptance testing to assure that production lots have met quality requirements. This may be accomplished either by 100% testing or by lot sampling, as is discussed in Chapter 5.

Storage, handling, and shipping may adversely affect the quality of food products (with the possible exception of some wines and cheeses). Lengthy storage permits settling of some emulsified liquids and may be responsible for damage from corrosion, infestation, or degradation. Damage from improper handling and shipping includes leaking or deformed containers, scuffed labels, or crumbled product. Abnormally high-temperature storage of "commercially sterile" canned goods may result in thermophilic bacterial spoilage. Temperature control during storage of frozen or refrigerated products may be critical. Control of the quality during this phase of product life starts with audits of shipping and storage procedures. Systems development for storage, handling, and shipping generally rest with the distribution manager, but quality problems associated with these areas can be highlighted by quality control personnel; perhaps some suggestions for quality improvement may be uncovered and made.

Product coding and case marking may seem like a simple housekeeping chore, but it can assume monumental proportions in the event of a product recall. Whether the problem involved in a recall is small (such as forgetting to include a minor ingredient in a short run or accidentally labeling the orange-flavored product *mandarin* orange-flavored) or a major product quality failure, proper coding can mean the difference between calling back a truckload or two from a small area or bringing back product from the entire distribution area for many days of production because the code could not be identified.

Government requirements for coding should be considered minimums. If possible, all information should be included in codes: date, shift, line and plant of origin, batch, and product description. Using the Julian Date Code (the number of the day in the year) and a combination of letters and colors, a great deal of information can be included in very little space. The month can be designated by using 1 through 9 for January through September and either *O, N, D* or inverted 1, 2, 3 for October through December. The year could be expressed as a single digit. The date portion of the code need not exceed five characters. Universal product codes, label edge scoring codes, and other ingenious schemes might also be used where they do not conflict with regulations.

Let us now consider production quality records. As these tend to become voluminous in a short time, procedures must be established for retaining only those records which are actually useful. The general rule is to retain records for no longer than the normal life of the product. This rule should be modified somewhat. To retain skeletal data for use in comparing product quality progress over several years and to retain records of prior formulations and methods in the event they may someday be used again, one scheme retains all records from one randomly selected day per month for a period of one year after the normal product life. Then it reduces older records by selecting at random one month per year for retention for two years. If quality records are retained in a computer, there is no reason to destroy any of them, because they may be stored in very little space on inexpensive disks. If records are on paper, however, a system such as the one described is recommended. Table 2-2 is a schematic representation of the records retention system for a product with a maximum life of two years.

The goals of a quality records retention system must be clearly established if they are to be effectively reached. These goals include:

1. Providing evidence that prescribed tests and inspections were made.
2. Identifying personnel responsible for performing quality tasks in case they need to be questioned.
3. Accumulating rejection, scrap, and rework data for calculations of quality costs (for cost reduction efforts).
4. Identifying product, tracing materials, equipment, operations, and date of production.
5. Establishing a data base for providing reports to management of quality performance, action taken, and recommendations for product, materials, process, or cost improvement.

As with any system, success is dependent on effective management. Unless the quality manager constantly reviews the goals of his programs and takes action when necessary, these goals tend to be overlooked, thereby dooming the quality function to that of a paperwork generator. This is no more apparent than in the

Table 2-2. A Records Retention System

Year	Complete file	Retain one day's files per month for year no.	Permanent file one day per year
1	1	—	—
2	1,2		
3	2,3	1	—
4	3,4	1,2	—
5	4,5	2,3	1
5	5,6	3,4	1,2
6	6,7	4,5	1,2,3
7	7,8	5,6	1,2,3,4

collection of quality records. They contain much valuable information but are of no use to the company unless they are studied, analyzed, and reported in digest form with recommendations for constant productivity improvement.

DEPARTMENTAL RELATIONSHIPS

A quality control system cannot function in a vacuum. It must be integrated into the entire fabric of the company. Unfortunately, there is no magic formula for accomplishing this. Cooperative relationships between departments cannot be expected to operate because of a presidential edict. There is little to be gained by writing such relationships into the quality manual. In fact, telling other departments how to function is bound to foster some antagonism within a company. The amount of antagonism could be significant or negligible—depending on the personalities of those involved. Here is the heart of the matter. A food processing company is often thought of as a steel and concrete building, filled with noisy equipment and fragrant vapors, fed by truckloads of raw material, and generating truckloads of finished product. It is much more than that. It functions only because there are people within it—with all their diverse personalities, needs, intellectual abilities, and goals. Men and women are all prone to jealousy and anger, pride and despair, energy and lassitude, ambition and despondence, and cooperation and selfish independence.

Is it possible to create a quality policy to lead all these individuals in one common direction? Perhaps so. If it is true that only management can create a system in which quality control works, then it seems logical to assume that a system can be developed in which all departments participate in contributing to quality control. The following oversimplified example might suggest a way.

Let us consider a company in which the purchasing department is expected to supply the production line with least-cost packaging material. The quality control department is expected to control the quality of the packaging at some predetermined level. Finally, the production department is expected to keep the packaging lines operating without interruption in order to minimize costs and eliminate a second shift or overtime. If purchasing supplies inexpensive but unreliable packaging, production will suffer numerous downtime episodes because of malfunction of the packaging material, quality control will reject vast quantities of substandard production, and production will have to schedule additional shifts to make up for lost product. Each department may have done its job in exemplary fashion, but the total effort was a costly quality failure. No amount of arguing between the departments will alleviate this problem. No department can be expected to tell another how to run its affairs without adding to the conflict.

Now let us bring enlightened management into the picture. If the company president decides to correct these costly problems, he might be able to arrange a solution which takes into account the special needs of individual departments. The purchasing, quality, production, and finance departments could meet once a week to select jointly a packaging supplier whose material cost and quality level would result in the highest output at the specification quality level and the lowest cost. This system of purchasing packaging materials overcomes differences in personalities and departmental goals. It is a company-wide, goal-oriented system and has a far better chance of success.

This problem and its solution have perhaps been oversimplified. For example, it might be worthwhile to include other departments, such as research or engineering, to consider alternative package design, production methods, or equipment that could use cheaper packaging material successfully. The principle of management systems is a valid one, however, and it is equally applicable to the production, shipping, commodity procurement, and possibly even personnel departments. It cannot completely overcome personality conflicts, but it can control them to a large extent.

This chapter is directed at the quality control department, not the president of the company. So far it has not supplied the quality manager with a solution to the problem of conflicting departmental goals. Teaching quality control methods to the company president can only be accomplished in small steps, as his priorities are usually directed elsewhere, but small steps can be effective. It would be a near miracle to provide a company with a new, effective, and complete quality control program in a single step. One can be developed in phases, however. If, in the preceding example, the purchasing manager were approached and shown how he might become a company hero by reducing the costs due to packaging problems on the line, then the first step toward interdepartmental quality operations would be launched. In a similar fashion, solutions to quality problems on the line which are discovered through analysis of quality reports can be discussed with managers in other departments so that they can take corrective action. This will assist in cementing relations with the quality department and in gradually developing the company-wide team approach to quality, frequently referred to as *total quality control.*

PRODUCT PERFORMANCE

A company's objective is to make a profit. Its objective in manufacturing a product is to satisfy consumer needs and expectations. Consumers provide the final— and most important—audit of product quality, and contact with consumers by the quality control department is the only way to really understand if product quality is satisfactory. Such contact may be direct—through form letters, coupons, mailings, or meetings—and it may be administered by outside consultants, internal sales and marketing departments, or by the quality control department.

Whenever practical, potential consumers of a proposed new product should be contacted to determine its intended use and practical requirements. Once a product is in the market, product information disseminated to the public must be carefully reviewed. It should reflect the needs of the consumer, as well as truthfully represent the product's performance, quality, and safety. Advertising should be reviewed for accuracy and conformance to regulations. Statements such as "twice the strength," "instantly soluble," "will keep forever in an unopened container," "sugar-free," "preferred by 80% of the product users," "organically produced," "reduced calories," and "guaranteed fresh" all require supportable evidence from the technical staff of either the research or quality control department. The quality system should provide the mechanics for reviewing marketing strategies where such advertising is proposed.

If nothing else, consumer contact in the form of consumer complaints must be handled quickly and carefully. It has been said that a single consumer complaint could represent 20 to 200 other consumers who had the same negative experience but merely changed brands rather than writing to the manufacturer. Every company has its own preferred procedures for responding to complaints, but in all cases, copies of complaints and responses should be forwarded to quality control so that efforts to prevent recurrences can be started immediately. Complaints fall into many different categories, depending on the product. They may be concerned with flavor, odor, foreign material, appearance, net volume or weight, presumed health effects, deterioration, and so on.

Complaints about a product that falls within specifications may be classified as nuisance complaints and handled by replacing the product for the consumer. When a product has apparently been produced outside of specifications, attempts should be made to obtain the package code date so that an immediate investigation can be performed. Keeping in mind the 20 to 200 unwritten complaints (see above), a single complaint could represent a serious quality problem—perhaps even a recall. Finally, when a product complaint indicates health hazards, all information should be immediately turned over to the company's insurance handler or legal department. No direct contact with the consumer by other company personnel is advisable because of possible legal liability considerations. However, close contact between the technical and legal departments is essential in order to establish the facts quickly.

Individual responses to consumer complaints are not the end of the story. Complaints should be compiled into categories by month and cumulatively for the past 12 months. (Do not use cumulative data for the current calendar year; this can be misleading.) Data should be listed by number of complaints and by percentages. If the company is unfortunate enough to have a large volume of complaints each month, control charts can be constructed to highlight both trends and emergencies. In any event, using percentage figures will highlight unusual situations. All data should be considered confidential and not distributed throughout the company except as required. A sample of such a compilation (from a fictitious company) is shown in Table 2-3.

An improvement over simple tabulation of complaints would be to include the sales figures as well as the complaints. Graphs showing types of complaints by product by month along with the number of units sold during that period may show that seasonal increases in complaints are accompanied (adjusting for lead time) by increases in sales. To further improve the interpretation of this chart, the

Table 2-3. Consumer Complaint Record

Complaint description	12-month cumulative		Jan.		Feb.		Mar.		Apr.	
	no.	%	no.	%	no.	%	no.	%	no.	%
Weak	82	25	5	17	6	14	3	8		
Too yellow	132	41	16	55	12	29	22	60		
Bitter	28	9	2	7	1	2	3	8		
Sticky	18	6	0	0	1	2	0	0	etc.	
Glass	1	—	0	0	0	0	0	0		
Dirt, sand	54	17	5	17	4	10	9	24		
Unsealed	7	2	1	4	18	43	0	0		
Total	322	100	29	100	42	100	37	100		

Note: Note the abnormal figure for unsealed in February.

data might be plotted as number of complaints per thousand (or hundred thousand, million, etc.) per month, which would deemphasize the seasonal nature of complaints. Simple computer programs are available which eliminate mathematical computations. It might be noted that seasonal complaints are not completely tied into seasonal sales data. It has been observed that in the colder and wetter parts of the sales areas, particularly during the winter months, the volume of complaints tend to rise over those from milder climates. This is probably due in part to some bored and perhaps irritable people, indoors and isolated by the weather, searching for amusement by writing letters of complaint for real and imagined product failures. Still, all complaints should be considered as legitimate until proven otherwise.

Another class of psychological complaint frequent occurs when a new product is introduced. Consumers with a negative attitude find change disagreeable and, although it is difficult to explain why they would try a new product in the first place, they may tend to complain about it because "it is different, therefore no good." Other consumers may be confused by a new or novel package, flavor, or product application. More often, however, the unfortunate truth is that there were still a few "bugs" in the new product which had not been discovered before the initial sales period. One can count on 100% inspection performed by the consumer to discover these failings. The more optimistic outlook regarding new product complaints is that, inevitably, the complaints per number of units sold decreases over time.

For those classes of complaints which generate fairly large numbers, it is possible to calculate the standard deviation and to construct three-sigma control charts to signal an alarm whenever an abnormally high number occurs. On the other hand, it would be far more logical to spend the time and effort to eliminate the causes for those classes of complaints, rather than to construct control charts.

CORRECTIVE ACTION

We shall define corrective action as creating a change in product development, production, or distribution to eliminate the risk of a quality failure. Note that repair or rework of a defect does not fall within this definition. Corrective action requires periodic assessment of quality failures—by product, by failure mode, and by material suppliers. The analysis of specific failures may require evaluating the quality system to prevent recurring failures.

Reviews of quality reports to management are the usual starting place for corrective action, but requests for such reports might be originated by any department or individual within the company. Under these circumstances, a request should be written, as this encourages the originator to think the problem through and provide sufficient information to launch an investigation. Even at this early stage, the projected cost of corrective action should be considered to determine the economic justification of further analysis and change.

Agreeing with Dr. Deming's principles of quality control, the *Food Processing Industry Quality System Guidelines* recommends designating responsibility for quality problems as management-controllable or operator-controllable (ASQC, 1986). If the operator is in control, it must be established that the operator knows what is expected of him or her, is able to determine how well he or she is conforming to those expectations, is able to adjust the process if he or she is not conforming, and is motivated to achieve what is expected. Management factors would include systems considerations related to design, materials, equipment, manufacturing, calibration and inspection methods, training, and personnel.

Statistical analysis of causative factors may lead to possible solutions. Once the important variables have been identified, appropriate controls should be established to provide manufacturing with the assurance that they can maintain those variables within established limits. The new controls must then be added to the list of audited tests and test data.

In the event of a serious or catastrophic quality failure, a withdrawal or recall of finished product must be considered; it is mandatory that the quality system include a series of procedures for responding quickly to such failures. Corrective action decisions regarding failures include evaluating all product that might be affected by the same quality problem. Production and distribution records should be available for determining where all affected product is stored—on the production floor, in the factory warehouse, in transit, in outside warehouses, in retail distribution systems, or in the hands of consumers. Procedures for locating and retrieving the suspect product should be at hand and implemented immediately after the necessary approval required by the company policy has been obtained.

A recall at any level is a serious event. Every food processing company is expected to have a recall system primed and ready to go at any time. Undetected "six-sigma events" that are beyond the control of management do occur: shattered mixer paddle, exploding light bulb with defective shield, incorrectly certified material, corroded valve seat, disintegrated contact sensor, atmospheric contamination, and even sabotage. This program, more than any other, must be formalized, documented, and kept up-to-date. Considerable assistance is available to set up such a program. The Code of Federal Regulations, Number 21, Part 7, covers FDA regulations. The Consumer Product Safety Commission, CPSC Act 1972, Section 15, also refers to recalls. The Grocery Manufacturers of America, Washington, DC, has published an excellent pamphlet (GMA, 1974), which describes a recall system for food products. Once a system has been established, it should be periodically tested by performing a mock recall. This will quickly verify that either the system is workable or that there are deficiencies in either the system or in the record-keeping procedures in various departments, or perhaps in outside warehouses. Such deficiencies must be remedied immediately and tested again with another mock recall.

Company policies differ regarding the desirability of written reports on quality failures such as recalls. Some companies may prefer that no written material be prepared; others use recall records as training materials for certain employees. It is suggested that those responsible for company legal matters be consulted on this matter. A concise report covering steps taken in the event of a recall may be required by the legal representative as defense material to be used should litigation arise as a result of product quality failure.

QUALITY PERSONNEL

In classical industrial engineering system studies, the four m's considered are men, materials, methods, and money. So far in this discussion of quality systems, we have only considered the last three. Even the best quality system, however, will not function without people who are carefully selected, trained, and motivated. Regardless of which department does the recruiting, quality control management is responsible for identifying the characteristics of a satisfactory quality employee. Quality control need not be staffed by the same type of outgoing and personable individual required by the sales department, and the qualifications should not call for a stereotypical unbending policelike person, either. Such character traits as dedication and honesty are required, along with alertness and a proficiency in at least basic mathematics. Mechanical aptitude is often useful as well. The employee recruitment and selection process varies from company to company, but final approval of a new employee should always remain with the quality manager. Vacancies may be filled by promoting from within, reviewing resume files, running newspaper and trade journal advertisements, enlisting the services of employment agencies, and other channels. The manager should have a list of qualifications available for guidance, and the final selection of employees should be based on capability and experience or the potential to qualify for the job.

Large corporations may have training departments to quickly indoctrinate new employees with company policies and procedures. In large or small companies, however, the quality manager is ultimately responsible for training new quality employees. Company policies and procedures manuals are useful for training newcomers to any department, but the training required for quality department employees is highly specialized. It must be general in nature and at the same time specifically oriented to the company's operations and quality policies. Perhaps the quickest (and most uniform) method of training quality control personnel is to have them study the quality manuals. Quality control personnel from other industries may not be aware of the critical nature of quality in the food industry. If a lawn mower is delivered with its blade's cutting edges installed backward, it can be repaired or returned for replacement. Not so with foods: If a food product contains a major defect which might affect the health of the purchaser, it cannot be repaired, and it might cause serious consequences. Many hardware industries have slow lines producing expensive products which are controlled by indirect nondestructive test methods. Most food processing plants operate high-speed processes and require different quality testing techniques. All quality employees should be required to take refresher courses geared to introducing new concepts and to the eradication of complacency and elimination of shortcuts.

Training of line and supervisory employees in the principles of quality performance is best handled by personnel other than those from quality control. Any production employee can see that putting the principles of quality workmanship into practice will make the quality control department look good. On the other hand, if the use of careful and precise quality methods reduces production rates, the production line supervisor might look bad. Of course, this is not necessarily true, but the problem exists nevertheless. By using trainers from other departments or outside the company, the possibility for such conflict is reduced. Obviously, the best control of quality performance is achieved when both production and quality control personnel are effectively trained in the principles of quality effort.

Motivation is another key to the success of quality control, but there is no simple formula for providing it. It is important for quality employees to understand the consequences of slipshod performance. It is equally important for line and supervisory personnel to be aware of this. A thorough knowledge of the principles of quality effort is required for quality control, but without the motivation to actually perform under these principles, the results will be less than acceptable. Many motivational tools are effective; unfortunately, most are short-lived. When used a second time, disinterest or even antagonism is often exhibited by employees. One motivational tool which seems effective in practically all situations is recognition of achievement. How to best reward achievement must be carefully thought out, however. A reward consisting of a dinner with the plant manager may work once or twice, but it might not even be considered a reward by many. It is also difficult to bring such a program to an end or replace it with another (to say nothing of the plant manager's long-term feelings about such an arrangement). Recognition can be as simple as a kind word from the company manager to an employee at his workstation. Financial awards, vacations, promotions, pronouncements at the Christmas party or in the company newspaper, plaques, gifts, posters—these are some of the many ways to recognize efforts and provide motivation. Some are more effective than others, of course, but whatever method is selected, it should be made clear to the recipient and his or her co-workers that the award is a recognition of outstanding performance—not a bribe to do the job correctly.

3   Control Charts

THE IMPORTANCE OF CHARTING

The ancient adage "A picture is worth a thousand words" is somewhat time worn but far from worn out. In a world where statistics and columns of numbers are not well understood, a picture of a process may be far easier to grasp than a quality summary laced with standards deviations, averages, equations, and numerical computations. The control charts used in statistical quality control represent a picture of a process. When used and continuously updated on the production floor, these charts represent a moving picture of the process.

The frequent misuse (or occasional fraudulent misrepresentation) of charts and graphs has placed them under suspicion. For example, suppose we are considering cost increases in changing raw product for a process from standard grade to choice. Preparing a report on this project, a simple bar chart (Fig. 3-1) may be used to compare the cost in 1988 and the cost in the year 2000 (when the new material is intended to be used). The bar on the left represents the 1988 cost. The bar on the right shows the cost in the year 2000—about double that of 1988.

To emphasize the difference, we may choose to represent the increase in cost

Figure 3-1.   Bar chart comparing costs for two years.

by a three-dimensional picture (Fig. 3-2). Here, as before, the vertical stack height of the year 2000 dollars is double that of the 1988 dollars, but by doubling the depth and width as well, the appearance of the new cost is dramatically (and untruthfully) magnified. The year 2000 dollars are not the same dollars as those of 1988.

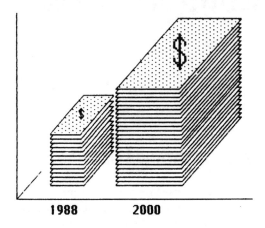

Figure 3-2. Three-dimensional diagram ex-
aggerates costs for year 2000.

Another trick commonly used to reach unwarranted conclusions by manipu-lating graphs is the magnifier principle, frequently practiced in newspapers to prove a point. Although it might be occasionally useful, extreme caution should be taken when employing this technique. For example, in Figure 3-3, the graph on the left shows the level of petty cash in an office safe over 15 years. The enlarged graph on the right has selected the most recent week, with grossly magnified axis values suggesting that petty cash is skyrocketing. In truth, it is most likely fluctu-ating in the general range where it has wandered over the past few years.

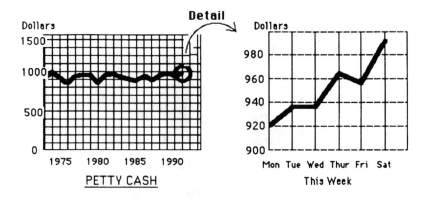

Figure 3-3.   Misleading effect of an enlarged portion of a chart.

Pie charts may be used to represent percentages or portions of a population and are sometimes useful. They are difficult to interpret if many sections are shown, and they can be misleading if they are portrayed in three-dimensional perspective. Note in Figure 3-4 how a 15% portion in the front of the pie can emphasize the profit of a proposal in the left pie slice; by contrast, note how the same 15% placed in the rear of the pie (and with a change in type size and font) suggests that the costs of this proposal are prohibitive compared to the meager profits.

THE PIE CHART GAME

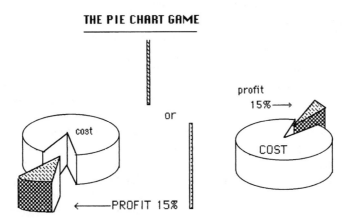

Figure 3-4.   Misleading effect of a pie chart.

On the other hand, charts and graphs are valuable tools for presenting statistical data. Of particular interest are the control charts used universally to present quality data (see Fig. 3-5). They are sufficiently simple to interpret so that misunderstandings are avoided. Regardless of type, control charts all contain a few fundamental characteristics. They contain upper and lower control limits within which all observations will lie if the process is in control, they contain a center line which is usually considered the target value for the process, and they generally show numbers along the vertical axis to define the values of the control limits and observations. Beyond these basics, a chart may be tailored to suit the requirements of the operation.

If one were to ask for the basis of statistical quality control, there might be con-

Figure 3-5.   Characteristics of a control chart.

siderable discussion, but the X-bar and R charts would rate very high on the list. These two charts are easy to prepare, simple to understand, and extremely useful in locating problems—sometimes even suggesting possible solutions. They are ideal tools for discovering ways to improve product quality and process control, and they can drastically reduce scrap and rework while assuring the production of only satisfactory product. They are used for controlling every step of a production process, for the acceptance or rejection of lots, for product improvement studies, and for early detection of equipment or process failures. Figure 3-6 is an X-bar control chart taken from an orange-flavored dry drink-mix packaging line at the filler discharge and shows the citrate concentration as determined periodically by the quality control laboratory (the average line has been omitted for clarity). At 9 o'clock, a point appears just above the control limit. Because, by chance alone, this may happen as infrequently as 3 times per 1000 observations and still represent a process in control, no action is indicated. The next few points are indeed within limits, and the process is allowed to proceed. At 12 o'clock, another point is out-of-limits; the recheck point immediately following shows, without doubt, that the process is far out of control. The line is stopped, and it is determined that an operator has inadvertently dumped a drum of lemon drink mix into the hopper during the noon shift change. The citrate control limits for the lemon mix product are 6–8 units, which would explain the increase. Additional evidence obtained by the laboratory confirms the mistake by product color analysis and flavor evaluation.

Figure 3-6. X-bar control chart for orange drink mix.

In addition to X-bar and R charts, a group of charts loosely defined as *attribute charts* are used for control of defect analysis. They are particularly useful for controlling raw material and finished product quality and for analyzing quality comments in consumer letters. An attribute is a characteristic of a product, a process, or any population that can be counted or tallied but cannot otherwise be described in incremental numbers. It would be more meaningful to define an attribute as a characteristic that is either satisfactory or unsatisfactory, go-or-no-go, defective or nondefective, good or bad, heavy or light, and so on. The only numbers that can be applied are the number or percentage of the satisfactory or unsatisfactory units. These charts (which are considered later in this chapter) are the p charts (fraction or percent defective), np charts (number of defectives), and c or u charts (number of defects). First, we explore the X-bar and R charts.

To start with, the *X*-bar and *R* charts are used for control of variables, measurements that are expressed in discrete numbers such as inches, pounds, pH units, angstroms, percent solids, or degrees centigrade. In the case of a leaking container, a variable measurement would be the rate at which gas flows through the leak in cubic centimeters per hour. An attribute measurement, on the other hand, would be the number of leakers or nonleakers in a batch or in some time period.

X-bar is usually written as $\bar{X}$ and is the average value of several measurements, each of which is called *X* and generally identified by means of subscripts (X_1, X_2, X_3, etc.). As we learn in Chapter 4, some distributions of data may be lopsided (skewed) because of peculiarities of the process from which the data have been obtained. It is also explained that taking averages of data from a process, regardless of the distribution of the data, tends to reduce these irregularities, resulting in a smooth distribution. A graph of a series of *X*-bars, therefore, shows a smoothed representation of the process being measured. This makes it easier to determine when the process is out of control, as smoothed data will lie in a relatively narrow range.

A possible weakness of an $\bar{X}$ chart is that individual points are represented by averages of data that might contain wide ranges of values but that are masked by the very smoothing effect that makes these charts so desirable. To overcome this difficulty, the range of data from which each average was obtained is also required and, in turn, its control values (upper and lower) must also be calculated.

PROCEDURE FOR CONSTRUCTING *X*-BAR AND *R* CHARTS

The preceding verbose description is a good argument for the statement that a picture is worth a thousand words. With that in mind, we proceed to construct an actual control chart for headspace in one-pound jelly bean jars packed on one of six lines in a plant. Our concern is that insufficient headspace is likely to crush the product when the lid is applied, and that excess headspace will allow the product to abrade during transport. Experience has shown that a headspace of about $\frac{3}{32}$ or $\frac{4}{32}$ of an inch is satisfactory, but plant operating people are not certain what any of the lines are capable of producing. We will work with Line No. 3 first. A preliminary chart can be prepared from about 100 headspace measurements. For a more realistic working chart, it is customary to obtain 30 subgroups of 5 samples for a total of 150. Refer to the data and calculations in Table 3-1 as we go through the following steps.

- When Line No. 3 is operating satisfactorily, select about 100 jars off the line before the lids are applied. All jars must be selected in sequence and kept in order until measured.
- Measure the headspace in each jar, in order, and record in groups of 5. Sample 1 measures $\frac{7}{32}$ and is recorded in the first column as 7. Sample 2 measures $\frac{10}{32}$ and is recorded in the second column as 10. Continue until reaching Sample 6, which measures $\frac{6}{32}$ and is recorded in the first column under the first sample taken.

Table 3-1.   _X_-Bar and _R_ Chart Calculations

Sample number						Average ($\overline{X}$)	Range (R)
1	2	3	4	5			
7	10	8	2	8		7	8
6	9	3	4	3		5	6
6	7	2	6	4		5	5
4	8	5	7	7		6.2	4
10	3	5	6	4		5.6	7
8	11	9	6	6		8	5
5	7	5	6	7		6	2
9	5	8	7	6		7	4
6	5	4	5	5		5	2
1	3	1	0	3		1.6	3
3	4	6	5	4		4.8	3
5	3	6	3	3		4	3
5	6	8	9	7		7	4
7	7	8	7	6		7	2
7	7	6	7	7		6.8	1
7	7	7	9	9		7.8	2
(add columns)					Totals	93.8	61
(divide by 16)					Averages 5.86		3.81

A_2 for sample of 5 = 0.577
D_4 for sample of 5 = 2.115 (from tables)
D_3 for sample of 5 = 0

Eventually, 16 rows of 5 columns will be completed. This totals 90 samples—close enough to the 100 generally required for a first attempt. The reason for the 5 columns will be explained shortly.

- Calculate the averages for each row. For example, 7 + 10 + 8 + 2 + 8 = 35. Divide 35 by 5 and record 7 under the AVERAGE heading.
- Record the ranges for each row under the RANGE heading. For example, in row 1 subtract 2 from 10 and list 8 under RANGE.
- Add all of the averages (93.8 total) and divide by the number of rows (16) to determine the grand average (5.86).
- Add all of the ranges (61 total) and divide by 16 to determine the average range (3.81).
- Refer to Table A-8 in the Appendix, "Factors for Computing Control Chart Limits," and find 0.577 under column A_2 for sample size of 5. Note that the values for samples of 2 through 4 are quite large and decrease in substantial quantities, but starting with a sample size of 5, these numbers decrease at a much smaller rate. This is the reason for combining samples in groups of 5.

If more precision is required at some later date, the test may be repeated with larger groupings and more samples to obtain 30 subgroups.
- Using the same table, find 2.115 under column D_4, and 0 for D_3. We have now prepared all the data necessary to calculate the control limits.
- Calculate the upper and lower control limits from the formulas as shown:

$$UCL_X = \overline{\overline{X}} + A_2\overline{R} = 5.86 + 0.577 \times 3.81 = 8.06$$

$$LCL_X = \overline{\overline{X}} - A_2\overline{R} = 5.86 - 0.577 \times 3.81 = 3.66$$

$$UCL_R = D_4\overline{R} = 2.11 \times 3.81 = 8.05$$

$$LCL_R = D_3\overline{R} = 0 \times 3.81 = 0$$

- Plot headspace against sample number, where the sample number refers to the average for each set of 16 sample groups.
- Draw the upper and lower control limits (8.06 and 3.66) and the grand average (5.86). This completes the X-bar control chart (Fig. 3-7).
- Plot headspace against sample number, where the sample number refers to the range for each subgroup of 5 samples.
- Draw the upper and lower control limits (8.05 and 0) and the average range (3.81). This completes the range chart (Fig. 3-8).

There are several points of interest in these charts. Note that in the control chart for averages (Fig. 3-7), even in this short burst of production, one of the points (number 10) is out of limits for headspace on the low side. Whether this is

Figure 3-7. X-bar control chart for jelly bean jars.

Figure 3-8. Range chart.

a desirable or undesirable measurement might be worth knowing and some effort should be made to find out what might be responsible for this apparently out-of-control jar. Perhaps the beans are oversized, there is less breakage, or the jar is overfilled. There is always the possibility that a peculiarity in one of the filling machine heads might explain the abnormality. In any event, if a cause is found, action should be taken to assure that this abnormality will not occur again (if undesirable) or to assure that it will always occur (if a lower headspace is desirable). The fact that the remainder of the measurements are in control should not be surprising. After all, the limits for control were defined by the data that are plotted.

One conclusion which may be drawn from this initial effort is that the "satisfactory" limits of $\frac{3}{32}$ to $\frac{4}{32}$, used informally in the past as specifications, are not attainable under this process as it now exists. The current process can run successfully between $\frac{3.6}{32}$ and $\frac{8}{32}$; if tighter specifications are truly important to the quality of the product, considerable work lies ahead to improve this process.

Having developed an initial control chart, the limits should be used every day for a while, plotting headspace averages of five samples taken off the production line every 30 minutes. It should not take long to find out if the limits are realistic. They should be refined as more data are gathered, until it can be decided whether process improvement is needed. Should that point be reached, simple improvements, which management is aware of but has been putting off, should be evaluated and installed. If improvement beyond this is required, a series of experiments should be designed to investigate other means of reaching the specification goals. To measure the effects of improvements, continue to plot the five sample averages; if a shift is suspected, prepare a new control chart showing the tighter control limits.

It might be noticed that the range control chart seems to have a downward trend. Because all of the values are within the control limits, it is probable that this unusual pattern has occurred by chance alone. On the other hand, it might be worth an initial investigation to find out if there is some special cause responsible for what appear to be more uniform weight ranges.

The chart for averages (Fig. 3-7) shows that the headspace is higher than it should be for optimum performance; the range chart (Fig. 3-8) shows more discouraging information. It indicates that, even though the average headspace could be reduced to a more acceptable level, the range between readings is still going to be very high. At the moment, the range is $\pm\frac{1}{32}$, and the present hoped-for specification limit is very much less, at $\pm\frac{0.5}{32}$. It now becomes obvious that there is work to be done to improve the performance of this line, starting with an attempt to narrow the wide range of headspaces currently produced. After that, the target level should also be reduced.

Subsequent chapters on process capability and process control contain detailed discussions on the meaning and use of quality control charts such as these.

PROCEDURES FOR CONSTRUCTING ATTRIBUTE CHARTS

There are four commonly used attribute charts, and each has a specific use. They are generally easier to construct and to use on a routine basis, although they occasionally lack the power of variable charts to spot problem areas quickly. A major advantage is the simple nature of the concept: Attribute charts are easily explained to line workers and management alike. Almost everyone understands the meaning of "percent defective" or "number of defects in the lot." The mathematics of attribute charts do not require tables. They can be applied to systems where measurements consist of pass–fail or where variable measurements are difficult or impossible to obtain. These are the charts:

- p Chart with Constant Lot Size: Used to determine control of percent defective units and to establish whether the process is in control for the day (week, month). Constant means within 20%.
- p Chart with Variable Lot Size: Usually intended to control percent defective units where the number of units varies from sample to sample. Determines if a process is in control for each lot's control limits.
- np Chart (also known as m Chart): Used to control the number of defective products in each lot and to assure that the process is in control. Requires constant lot size.
- c Chart: Used to determine if the number of defects in a single product is within control limits. Final inspection.

In Table 3-2, there are three classes of attributes considered: fraction defective (p), number of defectives (np), and number of defects (c, u). The mathematics involved in finding the averages is simply a matter of dividing the total number of nonacceptables by the number of samples. The calculation for control limits, based on this figure, is also simple. The formulas are shown in Table 3-3.

Table 3-2. Control Charts

Symbol	Name	General use	Application	Based on	Lot	Sample
$\bar{X}/R$	Shewhart	Locate assignable cause of process shift	Process control by variables (inches, ml, pH, ounces, count, etc.)	Normal distribution	Constant	Constant and arbitrary
p	Fraction or % defective, constant lot	Detect process change	Process control by attributes (leaks, flavor, blemishes, miscounts, seeds, dents, etc.)	Binomial distribution	Constant (within 20%)	Constant, 100% or sample
p	Fraction or % defective, variable lot	Detect process change	Process control by attributes	Binomial distribution	Varies	Varies 100%, average or stabilized
np	Number of defectives	Detect process change	Process control by attributes	Binomial distribution	Constant	Constant
c	Number of defects, constant sample	Accept or reject item	Process control by attributes	Poisson distribution	Constant or continuous	Constant
u	Number of defects, variable sample	Accept or reject item	Process control by attributes	Poisson distribution	Varies, or one unit	Varies

Table 3-3. Comparison of Attribute Charts

	p Charts	np Chart	c or u Chart
Chart name	Fraction or % defective	Number of defectives	Number of defects
Examine for	Defective unit	Defective unit	Defects in unit
No. of items in sample	n	n	n (usually 1)
Defectives in lot	np (or m)	np (or m)	—
Defects in lot	—	—	c
Average fraction defective	$\bar{p} = \dfrac{\sum m}{\sum n}$	$\bar{p} = \dfrac{\sum m}{\sum n}$	—
Average defectives	—	$\bar{m} = \dfrac{\sum m}{N}$	—
Average defects	—	—	$\bar{c} = \dfrac{c}{N}$
Control limits	$CL\ Y_p = \bar{p} \pm 3\sqrt{\dfrac{\bar{p}(1-\bar{p})}{n}}$	$CL\ Y_m = \bar{m} \pm 3\sqrt{\bar{m}(1-\bar{p})}$	$CL\ Y_c = \bar{c} \pm 3\sqrt{\bar{c}}$

Examples of the three basic types of charts selected from different industries illustrate the broad application of these techniques. The first example is the construction of a p chart for fraction defective.

Example
Percent Defectives Chart: p Chart with Constant Lot Size

In a plant producing drums of dill pickle chips, a day's production varies from 476 to 580 drums. Over the years, the major types of defective drums have been found to be leaks, contamination, discoloration, mislabeled drums, off flavor, overweight, and underweight. In all, there are 18 different types of defects which have been observed. The plant manager believes that, on the average, his operation produces no more than 2.5% defectives (or 0.025 fraction defective), based on the number of customer complaints he receives over the year. Using this meager evidence, he has set product specifications at 0–0.040 fraction defective (or 4% defective) drums. Because of a serious quality complaint from a major customer, a quality consultant is called in to find out if the specifications are being met. The consultant is charged with the responsibility of determining (a) if the specifications are realistic and (b) if the process is in control. If needed, the consultant is requested to make suggestions on how to improve quality and to maintain it at a higher level. The relatively few drums produced each day permit a test program in which the output is 100% inspected. If any defects are found in a drum, it is counted as a defective; if more than one defect is found in a single drum, it is still counted as a single defective. Data for 30 successive production days are presented in Table 3-4.

As shown by the data marked with solid circles (•), there are many drums (nearly one-third of the production) that equal or exceed the plant manager's arbitrary 4% defective specification limit. The consultant now proceeds to calculate the average fraction defective and average number of drums checked. In addition, the variation of sample sizes is also checked to see if a constant size chart will suffice.

$$\bar{p} = \frac{\Sigma m}{\Sigma n} = \frac{515}{15,565} = 0.033 \left(\text{or } 3.3\% \right)$$

$$\bar{n} = \frac{15,565}{30} = 519$$

$$\frac{580 - 475}{519} \times 100 = 20\% \text{ variation}$$

The average number of defective drums is not the 2.5% which the plant manager had believed, but is actually 3.3% during this 30-day run. These numbers (0.033 and 519) are all that is required to calculate the control limits for the process.

Table 3-4. Data for Defective Pickle Drums

Sample no.	Date (Oct.)	No. of drums (n)	Defectives (m)	Fraction defective (p)
1	4	502	18	0.036
2	5	530	13	0.025
3	8	480	13	0.027
4	9	510	15	0.029
5	10	540	21	0.039
6	11	520	17	0.040 •
7	12	580	28	0.048 •
8	15	475	10	0.021
9	16	570	23	0.040 •
10	17	520	16	0.031
11	18	510	15	0.029
12	19	536	22	0.041 •
13	22	515	18	0.035
14	23	480	12	0.025
15	24	548	24	0.044 •
16	25	500	11	0.022
17	26	515	19	0.037
18	29	520	16	0.031
19	30	485	13	0.027
20	31	520	14	0.027
21	Nov.	515	12	0.023
22	2	545	25	0.046 •
23	5	515	16	0.031
24	6	505	13	0.026
25	7	518	15	0.029
26	8	484	12	0.025
27	9	520	22	0.042 •
28	12	535	22	0.041 •
29	13	518	14	0.027
30	14	554	26	0.045 •
	Total	15565	515	

Upper control limit:

$$\bar{p} + Y_p = \bar{p} + 3\sqrt{\frac{\bar{p}(1-\bar{p})}{n}}$$

$$= 0.033 + 3\sqrt{\frac{0.033(1-0.033)}{519}}$$

$$= 0.033 + 0.023 = 0.056$$

Lower control limit:

$$\bar{p} - Y_p = 0.033 - 0.023 = 0.010$$

Expressing these findings in terms of percent defective (moving the decimal point two places to the right), the control limits are 1% defective to 5.6% defective, and the average is 3.3% defective. The consultant is now ready to plot the 30 days' data on a control chart.

Figure 3-9 shows that the process is within the statistical control limits, with the average number of defective drums each day ranging from about 2% to nearly 5%. The arrow points to the 4% maximum specification limit which the plant manager had optimistically hoped not to exceed. A more realistic specification would be 6% maximum defective drums. This chart more clearly shows the large number of drums exceeding the present specification than does the list of numbers in the inspection sheet (Table 3-4).

Having achieved an overall view of the plant's capabilities and finding it less than satisfactory, the next step is to attempt to improve the process. We will not follow this example to its conclusion, but will indicate the general direction the consultant will probably take. He will now pinpoint the causes for the defective drums, using a control chart for *defects* rather than the broader concept of *defectives* shown in Figure 3-9.

A chart for defects will show some periods where defects are very numerous, at or near the upper control limit. The causes should be immediately identified and new procedures installed to be certain that the problems are not repeated. Conversely, a defect chart also will likely show that on some days the plant is capable of producing fewer defects than others. By identifying the reasons for these improvements, additional procedure changes can be installed to ensure that the improvements will permanently become part of the operation.

Figure 3-9.   Fraction defective control chart, constant lot size.

Although the processing line is the first place to look for improvements, other areas which may be equally responsible include:

- Purchasing a better quality of raw materials.
- Working with the drum supplier to eliminate substandard drums.
- Engineering more gentle handling systems to reduce line damage.
- Identifying hazardous and critical control points so that rational inspection locations can be installed.
- Improving the training of processing personnel.
- Revising maintenance schedules so that equipment (scales, slicers, meters, mixers, thermometers, and so on) will function more reliably.
- Studying the quality drift cycles to precisely specify sampling schedules.

When all improvements are in place and the improvement rate in the defect chart is leveling off, it is time to return to the fraction defective control chart for averages and calculate the new process averages. Realistic specifications may now be written and quality control procedures revised to reflect the changes of the program. To continue the process at the new quality level, the fraction defective control chart should become a part of the daily process routine, with an immediate response to any adverse signal.

Example
Percent Defectives Chart: p Chart with Variable Lot Size

A small shrimp breading plant produces frozen shrimp in various types of packages. One of the first steps in the line is deveining the raw shrimp and packing them in intermediate cases for refrigeration and further processing. In an effort to control the deveining process, the plant manager attempts to use a p chart for attributes, but finds that the widely fluctuating lot sizes produce meaningless charts. A lot of 10 cases with 3 defectives calculates to 30% defectives; a lot of 40 cases with 3 defectives calculates to 7.5% defectives. These two figures make no sense when plotted on a graph with the same control limits. It becomes apparent that a variable lot size requires variable control limits. There seems to be no way to control the case output size, because they devein all raw shrimp received each day, and that quantity depends entirely on the luck of the local shrimp boats. To simplify the classification, an arbitrary system of defect quantities is devised. Any case containing more than 6 shells over $\frac{1}{16}$ inch, 4 tails, 10 veins over $\frac{1}{4}$ inch, or any of several other criteria is considered defective. Over a period of 10 days, the data for the number of cases produced and the number of defectives are tabulated in Table 3-5.

The first calculation determines the average fraction defective. (Note that this cannot be done until after the tenth day—one of the weaknesses of the attribute charting technique is that it is slow to set up initially and equally slow to evaluate the results of attempted process improvements.)

$$\bar{p} = \frac{\sum m}{\sum n}$$

$$= \frac{38}{420} = 0.0905$$

Table 3-5.   Data for Defective Shrimp Cases

Day	No. of cases (n)	Defectives (m)	Fraction defective (p)
1	40	4	0.100
2	30	2	0.067
3	60	4	0.067
4	60	7	0.117
5	50	5	0.100
6	40	2	0.050
7	10	3	0.300
8	50	6	0.120
9	40	3	0.075
10	40	2	0.050
Σ	420	38	

The upper control limit for the fraction defective is defined as

$$\text{CL } Y_p = \bar{p} + 3\sqrt{\frac{\bar{p}(1-\bar{p})}{n}}$$

$$= 0.0905 + 3\sqrt{\frac{0.0905(1-0.0905)}{n}}$$

$$= 0.0905 + 3\sqrt{0.0823n}$$

Because n varies each day, the calculation for the upper control limit is performed each day using this formula. The results are shown in Table 3-6 under the column heading, CONTROL LIMIT.

Plotting the fraction defectives for each day is performed in the same manner as for a constant lot size (see Fig. 3-9). The difference arises when the control limits are plotted. The lower limit is so close to zero that a lower control limit would not be particularly useful. The upper control limits are plotted for each sample size. Thus, the sample size for the first day is $n = 40$. The corresponding upper control limit is 0.227. Continuing, the control limits for each day are plotted along with that day's fraction defective until all 10 points are completed.

Interpreting the chart in Figure 3-10 discloses that, for the relatively small samples sizes, the control limits are very liberal, suggesting that the quality level has to be atrocious before a day's production is out of limits. As a matter of fact, the fraction defective average of 0.0905 (or 9%) is already an index of very poor quality performance; yet, even at this level, the operation appears to be in statis-

Table 3-6. Calculation of Control Limits

Day	No. of cases (n)	Defectives (m)	Fraction defective (p)	Control limit (UCL)
1	40	4	0.100	0.227
2	30	2	0.067	0.248
3	60	4	0.067	0.202
4	60	7	0.117	0.202
5	50	5	0.100	0.212
6	40	2	0.050	0.227
7	10	3	0.300	0.363
8	50	6	0.120	0.212
9	40	3	0.075	0.227
10	40	2	0.050	0.227

Figure 3-10. Fraction defective control chart, variable lot size.

tical control. As was the case with other control charts, when the sample size increases, the upper control limit changes less and less. This is clearly shown by comparing the distance on the vertical axis between $n = 10$ and $n = 20$ with the distance between $n = 50$ and $n = 60$. Most of the days sampled exhibited defective levels clustered near the 0.0905 average except for the spike data of Day 7. Again, even though this day showed a distressing level of 30% defectives, the numbers involved (3 out of 10 cases) are so small as to allow the control limits to indicate that the process is in control. As the sample sizes increase, this type of chart analysis becomes more powerful. For example, at $n = 500$, the UCL drops to about 0.13. This is quite close to the average value and provides a fairly tight control level.

The control chart for fraction defective with a variable lot size is cumbersome to use, difficult to interpret, slow to respond to changes, and clumsy to explain. Considering these difficulties, it is wise to avoid use of this chart whenever an alternative is available. There are two possibilities. The easiest way might be to change the sampling plan to a constant-size subgroup of 30 cases selected throughout the day's production, with 100% inspection on those days when too few total cases are produced. Another way would be to select a single major defect such as presence of veins and—in a subgroup of five cases per day selected at uniform intervals throughout the day—measure the length of veins in millimeters. From this, a variables control chart can be prepared to help locate the conditions producing an excessive number of these defects. As improved procedures are developed, they may continue to be monitored by control chart techniques, and the quality improvement program may be shifted to another type of defect.

Example
Number of Defectives Chart: np Chart (Also Known as m Chart)

An olive processing plant has a long-established quality control system using percent defective charts (p charts) to evaluate the quality level of production. However, they wish to start a quality improvement program and find that by grouping all defects into one category, specific quality problems are not highlighted. In the past, any jar containing blemished pieces, pit fragments, missing pimientos, leaking seal, low brine, or other defects was classified as a single defective. For this new program, the company has selected the presence of pit fragments as a major defect to be reduced or eliminated. A constant-size sample of 500 jars uniformly spaced within each production run shall be examined for pit fragments and an np chart for number of defective jars prepared.

Table 3-7 shows the first 10 days' data. Calculations based on these data establish the average number of defectives, the average fraction defective, and the control limit.

Average number of defectives:

$$\overline{m} = \frac{\Sigma m}{N} = \frac{129}{10} = 12.9$$

Average fraction defective:

$$\overline{m} = n\overline{p}, \qquad \overline{p} = \frac{\overline{m}}{n} = \frac{12.9}{500} = 0.026$$

Table 3-7. Data for Defective Olive Jars

Day	Sample size (n)	Number of defectives (m) or np
1	500	14
2	500	14
3	500	17
4	500	10
5	500	13
6	500	8
7	500	24
8	500	11
9	500	8
10	500	10
$N = 10$	$\Sigma\, n = 5000$	$\Sigma\, m = 129$

Control limits:

$$CL = \overline{m} \pm 3\sqrt{\overline{m}\left(1 - \overline{p}\right)}$$

$$= 12.9 \pm 3\sqrt{12.9\left(1 - 0.026\right)}$$

$$= 12.9 \pm 10.6$$

A control chart may now be constructed using these figures:

Average number of defectives = 12.9
Upper control limit = 12.9 + 10.6 = 23.5
Lower control limit = 12.9 – 10.6 = 2.3

This chart (Fig. 3-11) provides a good starting point for improving product quality insofar as pits are concerned. It shows that the pit level centers around 12.9%, and the quality level stays fairly close to that. On Day 7, some abnormality in the process caused the defects to rise out of limits on the high side, and operating conditions for that day should be examined closely for clues. Those familiar with the industry would know the most likely places to look—a change in raw material, maintenance programs on the pitter equipment, line speeds, and new line workers. If the "special cause" for this day's unusual production can be found, steps should be taken to ensure that it cannot happen again. After a few more days of production with this process modification, a new chart should be prepared with its associated closer control limits. Any time a point falls outside the limits, it should be analyzed to improve the process. At the same time, any observation which drops below the lower control limit should be studied with equal care; it, too, may indicate areas for quality improvement.

Figure 3-11. Number of defectives (*np*) control chart.

To see the effect of eliminating the problem responsible for the out-of-control level of Day 7, replace the 24 with 12.9, and calculate a new set of control points. The average would be reduced slightly to 11.8, and the upper control limit would now be 22.0. The significance of this little exercise is that it demonstrates how removal of just this one special cause of product quality variation can reduce the average defects by about 1% and the upper control limit by 1.5%. It should be obvious that continuing with this quality control chart and eliminating other "special causes" as they are discovered will appreciably reduce the defects caused by the presence of pits. Once the procedure reaches an equilibrium and further improvement appears unlikely, the next most important defect (perhaps blemishes) may be explored in a similar manner. An impatient manager might find this process painfully slow, but it can be a very effective quality improvement procedure.

Example
Number of Defects Chart—Constant-Size Sample: *c* Chart

In contrast to all of the charts where the quality issue is the number of *defectives,* the *c* chart is concerned only with the number of *defects.* The number of defects is measured for a subgroup of fixed size, and very often is concerned with a sample of size 1. The process to be measured may be a continuous one in which samples of a standardized number of units are selected at a prescribed constant interval, or a process producing a constant lot size may be checked with a standardized sample size. The sample of size 1 refers to a process in which the end product is a single sample—a batch of corn syrup, a side of beef, a vat of chocolate, or a bin of blended tea.

For our example, we shall select a batch of flavored syrup produced by a bottler of carbonated fruit drink. Ordinarily, a single batch of fruit-flavored syrup is produced each day to be diluted and packed on the following shift. The defects commonly found during quality control evaluations are foreign matter, temperature above 49°F, poor color, pH over 5.3, weak flavor, concentration incorrect (for each of 22 ingredients), incorrect viscosity, excessive bacteria (6 tests), yeast, or molds.

The data in Table 3-8 were obtained over 10 production days. Using these data, calculate control limits as follows:

$$\bar{c} = \frac{\Sigma c}{N} = \frac{201}{10} = 20.1$$

$$CL \ Y_c = \bar{c} \pm 3\sqrt{\bar{c}}$$

$$= 20.1 \pm 3\sqrt{20.1}$$

$$= \begin{cases} 20.1 + 13.4 = 33.5 \ \ UCL \\ 20.1 - 13.4 = 6.7 \ \ LCL \end{cases}$$

Again, we find one of the batches out of control in the initial chart. With the large number of attributes entered into the data, it is apparent that at least one of them is seriously out of control. If possible, the factor responsible should be isolated and studied with its own control chart until a reason for the variation has been isolated. Then new procedures should be initiated to prevent further occurrence of this defect in a day's batch. If the identity of the defect cause cannot be easily determined, control charts for small groups of defects should be prepared until the problem is discovered.

Example
Number of Defects Chart—Variable-Size Sample: u Chart
The u chart is a special case of the c chart and has limited use in the food processing industry. It is applied to products where the number of units in the sample subgroup varies, from a subgroup of one to perhaps a dozen units. The defects are fed into an equation to

Table 3-8. Data for Syrup Batch Defects

Sample number	Date (Feb.)	Number of defects (c)
1	3	15
2	4	21
3	5	27
4	6	18
5	7	39
6	10	10
7	11	18
8	12	22
9	13	13
10	14	18
		$\Sigma c = 201$

reduce the data to the number of nonconformities in a subgroup. The control limits are calculated in the same manner as for the c chart, except that the control limits are divided by the square root of the number of units in the subgroup. At best, the u chart serves as a score card to let management know how the quality was for a given period in the past. It is slow to react to quality control problems, and other types of control charts should be used before considering the u chart. Even in the automotive manufacturing industry, where the u chart is used in an attempt to control the number of defects in fenders or bumpers, it has limited effectiveness. It might be used as a blunt weapon to brandish at parts suppliers in an effort to assist them in product improvement, but it carries few, if any, suggested cures. Because of the crippling limitations of the u chart, no example is presented here.

It is difficult to generalize on the use and effectiveness of all types of attribute charts because there are so many possible applications in so many different food processing industries. Generally speaking, they are slow to react and tend to be constructed in such a way that they include a number of quality problems within one format. However, they do provide management with an easily understood picture of how quality has been controlled over some period of time. There are, of course, certain applications where precise variable measurements of a process are neither possible nor desirable, and attribute charts are quite satisfactory. Under some circumstances, they can suggest areas where a process should be modified to better control the quality level or, in some instances, actually improve the quality level. The one generalization which might be made is: Wherever possible, select a variable control chart instead of an attribute chart to control process quality.

This chapter has not covered all of the charts available for use in quality control applications. Many have specialized uses for specific industrial applications. Many are complicated and perhaps better understood by the statisticians who create them than by the plant operating and quality control personnel who must use them. Some mature industries tend to take shortcuts with the classical chart techniques discussed above, a somewhat risky practice. An example is the abandonment of range charts when experience has shown that an averages chart is sufficiently indicative of process quality. Sooner or later, one will probably miss an out-of-control production period when, for example, a worn machine abruptly spews out widely varying product that somehow continues to average out within the control limits. This is not to suggest that some experimentation with the tried and true control chart techniques should not be attempted. But it is strongly recommended that experimentation be delayed until all plant processes have been exposed to the simpler procedures long enough to ensure that the processes are under excellent control, and further quality improvements using these principles are unlikely to be discovered.

4   Fundamentals

There are four distributions of data that form the basis for statistical quality control: binomial, Poisson, normal, and Student's *t*. Common to these distributions are the concepts of measures of central tendency and measures of dispersion.

ANALYSIS OF DATA

Measures of Central Tendency

In order to illustrate a large quantity of data (population) with a single number, perhaps the best single measure is the arithmetic mean. This is simply the total of all values, divided by the number of items.

$$\text{mean } (\mu) = \frac{X_1 + X_2 + X_3 + \cdots + X_N}{N} = \frac{\Sigma X}{N}$$

Another illustration of the data is called the median. When the numbers are arranged in order of magnitude, the median of the set of data is the middle number if *N* is odd and the average of the two middle numbers if *N* is even. For example,

$$2, 3, 7, 9, 21, \mathbf{22}, 45, 67, 68, 69, 72$$

$$\text{median} = 22$$

$$4, 6, 9, 12, 24, \mathbf{45, 56}, 58, 59, 63, 68, 69$$

$$\text{median} = (45 + 56)/2 = 50.5$$

As seen in these examples, neither the mean nor the median has to be one of the numbers observed in the set of data.

Tables 4-1 and 4-2 demonstrate calculation of the population mean by different methods. Both examples are based on the same data obtained by weighing the net contents of 184-gram jars of pistachios taken in sequence from a production line. The data have been arbitrarily arranged in groups of five weighings. The first method for calculating the mean is based on the actual data; the second method uses deviations from the target weight of 184 grams. Both produce the same mean. The mean of the population is known as μ; the mean of the samples representing the population is $\bar{X}$.

Measures of Dispersion

The mean and the median do not provide much information about the distribution of data. Two different sets of data may have the same means, but the dispersion of data about the two means may be totally different. The dispersion is better described by the terms *range, variance,* and *standard deviation.*

The range (R) is found simply by subtracting the smallest value from the largest value found in the data. By itself, the range has limited use. This is because the smallest or the largest value may be extremely far from the mean value, thus giving an exaggerated impression of the dispersion of data.

Table 4-1. Data for Pistachio Net Weights

Sample	X_1	X_2	X_3	X_4	X_5	Total	Mean
1	187	190	183	185	188	933	186.6
2	186	182	183	185	189	925	185
3	187	189	192	192	188	948	189.6
4	190	189	189	187	187	942	188.4
5	188	188	183	187	183	929	185.8
6	187	185	187	190	185	934	186.8
7	194	189	194	186	187	950	190
8	183	188	187	187	183	928	185.6
9	189	192	194	183	188	946	189.2
10	188	186	193	191	185	943	188.6
11	187	187	188	182	183	927	185.4
12	188	188	190	190	183	939	187.8
13	186	189	194	186	192	947	189.4
14	189	184	194	188	190	945	189
15	190	186	189	191	187	943	188.6
16	188	189	185	185	183	930	186
17	185	188	193	188	188	942	188.4
18	187	187	190	189	183	936	187.2
19	194	183	191	191	192	951	190.2
20	186	188	188	180	186	928	185.6
						$\Sigma X = 18766$	$\bar{X} = 187.66$

Table 4-2. Deviations of Pistachio Net Weights

Sample	X_1	X_2	X_3	X_4	X_5	Total	Mean	Range
1	3	6	−1	1	4	13	2.6	7
2	2	−2	−1	1	5	5	1	7
3	3	5	8	8	4	28	5.6	5
4	6	5	5	3	3	22	4.4	3
5	4	4	−1	3	−1	9	1.8	5
6	3	1	3	6	1	14	2.8	5
7	10	5	10	2	3	30	6	8
8	−1	4	3	3	−1	8	1.6	5
9	5	8	10	−1	4	26	5.2	11
10	4	2	9	7	1	23	4.6	8
11	3	3	4	−2	−1	7	1.4	6
12	4	4	6	6	−1	19	3.8	7
13	2	5	10	2	8	27	5.4	8
14	5	0	10	4	6	25	5	10
15	6	2	5	7	3	23	4.6	5
16	4	5	1	1	−1	10	2	6
17	1	4	9	4	4	22	4.4	8
18	3	3	6	5	−1	16	3.2	7
19	10	−1	7	7	8	31	6.2	11
20	2	4	4	−4	2	8	1.6	8
						366	73.2	140

Note: $\overline{X}$ = 184 + (366/100) = 187.66.

For this reason, the variance (σ^2) is more commonly used to measure dispersion of data. The variance is calculated by subtracting each item (X) in the distribution from the mean μ, squaring each of the resulting figures, adding the results, and dividing by the number of items (N). Note that if the subtraction operation were not squared, the final figure would be zero, because half of the values are positive and half of the values are negative with respect to the mean. Another way to express this is to state that the average of all of the deviations from the mean will always turn out to be zero. By squaring, all of the figures become positive.

Unfortunately, the variance is expressed as the square of some number, and has limited direct use in quality control. For example, if all of the observations were weights of a distribution of packages, the variance calculated in the preceding example would be expressed in (grams)^2. To eliminate this complication, the square root of the variance is commonly determined, and the resulting figure is referred to as the standard deviation (or σ).

When using this formula to determine variance and standard deviation, there are many opportunities for calculation errors because of the many subtractions required. To avoid such errors, a simpler formula may be used in which only a single subtraction is needed:

$$\sigma = \sqrt{\frac{\Sigma(X - \bar{X})^2}{N}}$$

When N is large, it is not always possible or desirable to calculate μ and σ from the entire population, and estimates of $\bar{X}$ (the sample mean) and s (the sample standard deviation estimate) obtained from a sample of n items ($n < N$) from the population are used to approximate μ and σ. These estimates are calculated as follows:

$$\bar{X} = \frac{X_1 + X_2 + X_3 + \cdots + X_n}{n}$$

and

$$s = \sqrt{\frac{\Sigma(X - \bar{X})^2}{n-1}} = \sqrt{\frac{\Sigma X^2 - \frac{(\Sigma X)^2}{n}}{n-1}}$$

Note that $\bar{X}$ is generally close to the population mean, μ. Similarly, s is close to the value of the population standard deviation, σ.

When working with frequency distributions, the formulas and calculations are quite similar. Using the same data as in the pistachio examples, note the similarity of computations:

$$\bar{X} = \frac{fX_1 + fX_2 + fX_3 + \cdots + f_nX_n}{n} = \frac{\Sigma fX}{\Sigma f}$$

and

$$s = \sqrt{\frac{\Sigma f(X - \bar{X})^2}{n-1}} = \sqrt{\frac{\Sigma fX^2 - \frac{(\Sigma fX)^2}{n}}{n-1}}$$

where $n = \Sigma f$;

$$\bar{X} = \frac{\Sigma fX}{\Sigma f} = \frac{18,766}{100} = 187.66$$

$$s = \sqrt{\frac{\Sigma f(X - \bar{X})^2}{n-1}} = \sqrt{\frac{984.44}{99}} = 3.15$$

The FREQUENCY columns in Table 4-3 are more usually assembled vertically in a format referred to as a histogram. The same data have been rearranged as shown in Figure 4-1 and Table 4-4.

$$\bar{X} = 184 + \frac{366}{100} = 187.66$$

$$s = \sqrt{\frac{\sum fX^2 - \frac{(\sum fX)^2}{n}}{n-1}} = \sqrt{\frac{2324 - \frac{(366)^2}{100}}{99}} = 3.15$$

PROBABILITY

Statistics and probability are so fundamentally interrelated that it is impossible to discuss statistics without an understanding of the meaning of probability. Probability theory makes it possible to interpret statistical results. Many statistical procedures involve conclusions that are based on samples, and these are always affected by random variation. Probability theory allows us to express numerically the inevitable uncertainties in drawing these conclusions.

When assessing the quality of a lot of goods, rarely is the total lot examined

Table 4-3. **Frequency and Standard Deviation Calculations for Pistachios**

X	Frequency	f	fX	$X - \bar{X}$	$f(X - \bar{X})^2$
180	/	1	180	−7.66	58.6756
181		0	0	−6.66	0
182	//	2	364	−5.66	64.0712
183	/////////////	12	2196	−4.66	260.5872
184	/	1	184	−3.66	13.3956
185	////////	8	1480	−2.66	56.6048
186	////////	8	1488	−1.66	22.0448
187	///////////////	15	2805	−0.66	6.5340
188	/////////////////	17	3196	0.34	1.9652
189	///////////	11	2079	1.34	19.7516
190	////////	8	1520	2.34	43.8048
191	////	4	764	3.34	44.6224
192	/////	5	960	4.34	94.1780
193	//	2	386	5.34	57.0312
194	differing	6	1164	6.34	241.1736
		100	18766		984.44

Note: Range of weights = 194 − 180.

Figure 4-1. Histogram for distribution of weights.

Table 4-4.   Range of Differences

X − 184	f	fX	fX²
−4	1	−4	16
−6	0	0	0
−2	2	−4	8
−1	12	−12	12
0	1	0	0
1	8	8	8
2	8	16	32
3	15	45	135
4	17	68	272
5	11	55	275
6	8	48	288
7	4	28	196
8	5	40	320
9	2	18	162
10	6	60	600
	100	366	2324

piece by piece. The costs associated with such a practice generally make it impractical, and because of inaccuracies, 100% inspection is rarely 100% effective. Inaccuracies may be caused by limits of the test procedures or boredom of the inspectors (often bordering on hypnosis), which allows defects to pass unnoticed. Instead, a sample of the lot is usually tested and the number of defective articles recorded. For this result to be meaningful, the sample should be representative of the lot and selected in such a way that each article within the lot has an equal chance of being selected. This is referred to as a *random sample.*

Within a lot, if there are *r* defective articles in a random sample of *n* articles, then in the selection of each article from the lot, the probability of obtaining a defective article is $p = r/n$, and the probability of obtaining a nondefective article is $q = (n - r)/n$. If $p = 0$, there are no defective articles. If $p = 1$, all articles are defective.

For example, if $\frac{1}{10}$ (or 10%) of the articles in a lot are defective, then $\frac{9}{10}$ (or 90%) are nondefective. This may also be expressed as $p = \frac{1}{10}$ and $q = \frac{9}{10}$. Frequently, defects are reported in percentages: A probability of $\frac{5}{100} = 0.05$ may be referred to as a 5% probability.

Another simple concept of probability frequency used in procedures for chart control of quality is the expression np. If p is the probability of a defective article in a single trial, then the expected number of defective articles in n trials is np. For example, if there are 15 defective articles in a lot of 100, then in a random selection of a single article from the lot, the probability of drawing a defective article is $p = \frac{15}{100} = 0.15$. In drawing 20 articles from the lot, one might expect to obtain $np = 20(0.15) = 3$ defective articles. Note that these calculations refer to the long run. It is entirely possible that a sample of 20 might contain no defectives at all, but "in the long run," the probability is that 3 defective articles would be found in each sample of 20.

In the course of controlling quality, observations of one type or another are required. These observations may be physical measurements of length, weight, volume, wavelength, voltage, and so on; or the observations might be merely counts, such as the number of products which were incorrect, too big, too small, scratched, dented, and so on. The observations may be made over a period of time (a day, an hour, etc.), within a selected quantity (a production lot, a 30-gallon tank, a pallet load, etc.), until some required event occurs (failure of an electrical device, end of shelf life, melting point temperature, a satisfactory item is produced, etc.), or any other arbitrarily established limits.

Observations within the limits selected are referred to as *data*. Data may be obtained from selected items (*samples*) within those limits or from all of the items (the *population*). The position, arrangement, or frequency of occurrence of the data within the population is defined as the *distribution* of data.

Depending on the product, process, or system being examined, data may fall within many different types of distributions, and analysis of the data cannot logically proceed without first understanding which distribution is present. Following are brief descriptions of distributions commonly encountered in quality control calculations.

- Binomial—The probability that an event will happen exactly x times in n trials is calculated from a formula based on the Binomial Theorem, $(a + b)^n$ (also referred to as the Bernoulli distribution). The data conforming to these conditions define the resulting binomial distribution curve, which is useful in determining sampling plans and establishing control charts for defectives.
- Poisson—An approximation to the binomial that is valid when the samples are large and probabilities are small. Used for developing defect control charts and for calculating sampling plan probabilities.
- Normal—A common distribution found when the variable examined is the result of many causes that have a 50/50 chance of occurring. Average and

range control charts are based on this curve. Although the population may not be normally distributed, the averages of groups of samples selected from it generally follow a normal distribution.

- t-Distribution—A somewhat non-normal curve produced when comparing samples and population means when the population standard deviations are unknown and must be estimated from the samples.
- F-Distribution—The distribution of the ratio of two estimates of variance.
- Exponential—Logarithmic curves that describe events such as flavor loss, shelf life, or container failure.
- Weibull—These may take many shapes and are based on a single formula with three variables: shape, scale, and location. Used to study shelf life and product failure rates.
- Chi square—As contrasted with t-distribution, the chi-square distribution is used when the standard deviations are known for the sample and for the population. These are then compared.
- Others—Multinomial (a generalization of the binomial), hypergeometric (samples removed from population for testing are not replaced), uniform, Cauchy, gamma, beta, bivariate normal, geometric, Pascal's, and Maxwell.

BINOMIAL DISTRIBUTION

Many quality control problems deal with a population in which a proportion p of the individuals has a certain characteristic and a proportion, $q = 1 - p$, does not. This is called a binomial population because each individual in it falls into one of only two classes. Sampling for defective articles in a lot where each article selected at random is either good or defective leads to a binomial distribution.

In a random selection of a sample size n from a large lot, if p is the probability of obtaining a defective item in each draw and $q = 1 - p$ is the probability of obtaining a nondefective item, then the probability $P(r)$ that there will be exactly r defective items in the sample size n is the term in the binomial expansion of $(q + p)^n$ for which the exponent of p is r; that is,

$$P(r) = \binom{n}{r} q^{n-r} p^r$$

where

$$\binom{n}{r} = \frac{n!}{(n-r)! \, r!}$$

[Values of $\binom{n}{r}$ may be found in Table A-4 in the Appendix.]

A distribution whose terms are proportional to successive terms of the binomial expansion is called a binomial distribution. Its mean is $\mu = np$ and its standard deviation $\sigma = \sqrt{npq}$.

A binomial probability distribution may be represented by a histogram. This is a figure obtained by plotting all possible values of r along a base line, and erecting on each an equal-base rectangle whose height (and area) is proportional to the corresponding probability.

Example
In a manufacturing process in which it is known that 10% of the manufactured articles are defective, a sample of five articles is taken. What is the probability that in the sample there are (a) exactly 0, 1, 2, 3, 4, and 5 defective articles, and (b) 2 or fewer defectives?
(a)

No. of defectives (r)	Probabilities $P(r)$			
0	$P(r=0) = \binom{5}{0}\left(\frac{9}{10}\right)^5\left(\frac{1}{10}\right)^0$	$=$	$\left(\frac{9}{10}\right)^5$	$= \quad 0.59049$
1	$P(r=1) = \binom{5}{1}\left(\frac{9}{10}\right)^4\left(\frac{1}{10}\right)^1$	$=$		0.32805
2	$P(r=2) = \binom{5}{2}\left(\frac{9}{10}\right)^3\left(\frac{1}{10}\right)^2$	$=$		0.07290
3	$P(r=3) = \binom{5}{3}\left(\frac{9}{10}\right)^2\left(\frac{1}{10}\right)^3$	$=$		0.00810
4	$P(r=4) = \binom{5}{4}\left(\frac{9}{10}\right)^1\left(\frac{1}{10}\right)^4$	$=$		0.00045
5	$P(r=5) = \binom{5}{5}\left(\frac{9}{10}\right)^0\left(\frac{1}{10}\right)^5$	$=$	$\left(\frac{1}{10}\right)^5$	$= \quad \dfrac{0.00001}{1.00000}$

(b)

$$P(r < 3) = P(r = 0) + P(r = 1) + P(r = 2) = 0.99144$$

Under the conditions of the preceding example, in repetitions of the sampling, one would expect to obtain a sample without defects more than one-half of the time, with 1 defect about one-third of the time and 2 defects about 7 times in 100 samplings. (See Fig. 4-2.)

Probability

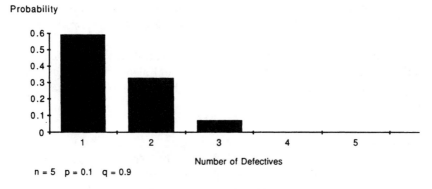

n = 5   p = 0.1   q = 0.9

Figure 4-2.   Histogram of binomial distribution example.

Figure 4-3.   Histogram for $p = q = 0.5$.

In the event that the values of p and q approach equality (for example, $p = q = 0.5$), the histogram for any value of n approaches symmetry around the vertical line erected at the mean. The total area included in the blocks of the histogram always equals 1. (See Fig. 4-3.)

POISSON DISTRIBUTION

An approximation to the binomial distribution, the Poisson distribution, may be used when (a) the probability of defectives is small (less than 0.1%), (b) the number of observations is large (more than 16), and (c) the sample is small compared to the population (< 10%). The Poisson distribution may be used to prepare control charts for defects, but perhaps has its greatest use in calculating probabilities for developing sampling plans. Much of the mathematics has been prepared in the form of a table which simplifies its use. Table A-1 in the Appendix gives the sums of the probabilities of c or fewer defects for various values of $np' = c'$.

If n is large and p and q are small, so that np is fairly small, that is,

$$\mu = np \leq 5$$

then the binomial probability

$$P(r) = \binom{n}{r} q^{n-r} p^r$$

is given approximately by the equation

$$P(r = c) = e^{-np} \frac{(np)^r}{r!} \quad (r = c = 0, 1, 2, \ldots, n \quad \text{and} \quad e = 2.178 \ldots)$$

A distribution where the frequencies are given by this formula is called a Poisson distribution and is extremely useful in the inspection and control of manufactured goods where the proportion of defective articles p' in a large lot is expected to be small. Following are several examples.

Example 1
In a large shipment of articles, 5% are known defective. What is the probability that, in a sample of 100 articles from the shipment, there are 2 or fewer defectives?

$$\mu = np' = 100(0.05) = 5 \quad c = 2$$

No. of Defectives

c np	0	1	2	3	4	5	6	7	8
Mean 5.0	.007	.040	.125	.265	.440	.616	.762	.867	.932

Therefore, $p(c = 2) = 0.125$ is the probability for 2 or fewer defectives; that is, the chances are between 12% and 13% that a sample of 100 will contain 2 or fewer defectives.

Example 2
If the true process average of defectives (proportion defective) is known to be 0.01, find the probability that in a sample of 100 articles there are (a) no defectives, (b) exactly 2 defectives, (c) fewer than 5 defectives, and (d) at least 5 defectives (i.e., 5 or more). Here, $\mu = np' = 100(0.01) = 1$; c is the number of defectives less than and including c. (It normally is the acceptance number, that is, the maximum allowable number of defects in a sample of size n to be acceptable.) We use the table to find probabilities for various values of c.

(a) $P(c = 0) = 0.368$.
(b) $P[(c = 2) - (c = 1)] = 0.920 - 0.736 = 0.184$.
(c) $P(c = 4) = 0.996$.
(d) $P[1 - (c = 4)] = 1 - 0.996 = 0.004$ (total of all terms = 1).

Example 3
A pump in a cannery has a failure, on the average, once in every 5000 hours of operation. What is the probability that (a) more than one failure will occur in a 1000-hour period, and (b) no failure will occur in 10,000 hours of operation?

(a) $np' = 1000(\frac{1}{5000}) = \frac{1}{5} = 0.20$. Because the table gives the probability of one or less failures at $c = 1$, then the probability of more than one failure is represented by $P[1 - (c = 1)]$ because the total of all terms equals 1. From the table,

$$P[1 - (c = 1)] = 1 - 0.982 = 0.018$$

(b) $np' = 10,000(\frac{1}{5000}) = 2$; $c = 0$. From the table,

$$P(c = 0) = 0.135$$

Example 4

A purchaser will accept a large shipment of articles if, in a sample of 1000 articles taken at random from the shipment, there are at most 10 defective articles. If the entire shipment is 0.5% defective, what is the probability that the shipment will be accepted?

$np' = 1000(0.005) = 5$; $c = 10$. From the table,

$$P(c = 10) = 0.986$$

Therefore, the shipment has a 98.6% probability of being accepted.

THE NORMAL DISTRIBUTION

In quality control, the normal distribution has the widest application of any of the distributions. In binomial and Poisson distribution, the random variable takes only a countable (or discrete) number of possible values. By contrast, many observations involve a process of measurement and in all such processes, we are dealing with quantities that have an infinite number of gradations. This gives rise to the normal distribution: As the number of gradations (n) approaches infinity, the resulting histogram approaches a curve. (See Figs. 4-4 and 4-5.) Early studies of the weights of groups of people, lengths of corn stalks, and other naturally occurring phenomena produced distributions approximating bell-shaped curves. Because these curves are found so often in nature, they are called *normal curves*.

A clingstone peach cannery may have a target drained weight of 19 ounces in a 2½ size can. If the filling operation is in control, most of the production will center near the target weight, but the frequency will taper off from the average as the deviations become wider. The need to establish attainable control weight limits on either side of the target weight is obvious. Consumers feel cheated if a product weighs less than stated (and this could be in violation of the law). and overly excessive weights can be costly to the producer. The normal distribution is used to calculate the values of less-than-stated and overly excessive weights.

The equation which defines the normal probability is

$$P = \frac{1}{\sigma\sqrt{2\pi}} e^{-(X-\mu)^2/2\sigma^2}$$

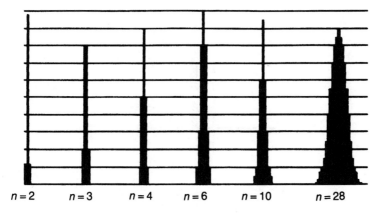

$n = 2$ $n = 3$ $n = 4$ $n = 6$ $n = 10$ $n = 28$

Figure 4-4.   Normal distribution as n increases.

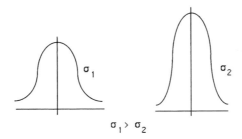

σ_1

σ_2

$\sigma_1 > \sigma_2$

Figure 4-5.   Frequency distributions with
different σ values.

where P is the relative frequency (or probability) with which a value of X occurs and μ and σ are the mean and standard deviation, respectively ($\pi = 3.14159$ and $e = 2.178$). The value of σ determines the concentration of data about the mean, as shown in Figure 4-5.

In order to understand the power of the normal curve and its application to quality control, let us consider some of its universal qualities. The areas under the normal curve represent probabilities of the normal distribution. As was the case of the histograms depicting the binomial distribution, the total area under the normal curve equals 1. Therefore, areas under the curve may be considered as proportions, probabilities, and (when multiplied by 100) as percents.

In Figure 4-6, the area A under the normal curve between two vertical lines at X_1 and X_2 represents the probability that a randomly drawn measurement X will fall between X_1 and X_2. In algebraic terms, $P(X_1 \geq X \geq X_2) = A$.

To simplify calculations, it is customary to use tables to find the areas under the normal curve. Because there is an unlimited number of normal curves, the tables were constructed for universal use as follows: The deviation of any measurement X from the mean is expressed in terms of standard deviations:

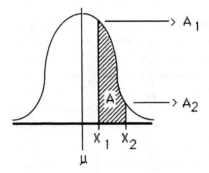

Figure 4-6. Area of probability between X_1 and X_2.

$$Z = \frac{X - \mu}{\sigma}$$

Referring to Figure 4-6,

$$A_1 - A_2 = P(X_1 \leq X \leq X_2) = A$$

and

$$Z_1 = \frac{X_1 - \mu}{\sigma} \qquad Z_2 = \frac{X_2 - \mu}{\sigma}$$

These Z equations change any normal measurement to another with zero mean and unit standard deviation. This transformation to normal deviate Z makes it possible for one table to suffice for all normal distributions.

Table A-2 in the Appendix is a compilation of areas under the normal curve to the right of values of Z (or to the left of values of $-Z$). Certain special areas are shown in Figure 4-7. These areas are designated by horizontal axis divisions expressed in terms of the two scales, $X - \mu$ and Z.

Distribution of Sample Means

In discussions of normal distributions, many references are made to *parameters* and *statistics*. The following definitions should make the differences between these terms clear.

Parameters are the values of μ (the mean) and σ (the standard deviation) obtained from the *population*.

Statistics are the values of $\bar{X}$ and s, the estimates of μ (the mean) and σ (the standard deviation), obtained from *samples*.

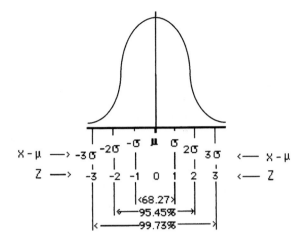

Figure 4-7. Areas under the normal curve in terms of μ, σ, z, and %.

The relationship between a population and samples taken from that population is vital to establishing quality control chart methods. If measurements in a population are distributed normally, then averages of groups of measurements from that population will be found to be closer to the mean than the original measurements. And, more important, if measurements of a population are *not* distributed normally, averages of groups of measurements from that population *will* approach a normal distribution.

This may be expressed in mathematical terms. When a lot or population is sampled, the sample mean is usually denoted as $\overline{X}$. Sampling generates a population of $\overline{X}$'s with a mean $\mu_{\overline{X}}$ and standard deviation $\sigma_{\overline{X}}$ of its own. If the population consists of k sample means, $\overline{X}$, it can be shown that

$$\mu_{\overline{X}} = \frac{\sum \overline{X}}{k} = \mu \qquad \sigma_{\overline{X}} = \sqrt{\frac{\sum \left(\overline{X} - \mu \right)^2}{k}} = \frac{\sigma}{\sqrt{n}}$$

In other words, if a set of measurements X has a distribution with mean μ and a standard deviation σ, and if all possible samples of n measurements are drawn, then the sample mean $\overline{X}$ will have a distribution which, for larger and larger n, approaches the normal distribution with mean μ and standard deviation $\sigma/\sqrt{n}$. This is known as the *central limit theorem* and states that for significantly large n (at least 30), $\overline{X}$ will be approximately normally distributed, even if the original population of X does not satisfy a normal distribution.

Figure 4-8 demonstrates the way in which a population distribution is narrowed as the sample size is increased from 4 to 16 measurements. Three curves are shown: the distribution of the population from which the samples are drawn,

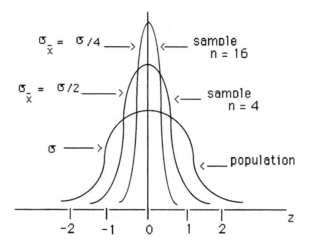

Figure 4-8. Comparison of population curve with curves for two sample sizes.

the distribution of sample means when samples of size 4 are selected, and the distribution of sample means when samples of size 16 are selected. With increasing *n*, the curve becomes taller and narrower. The areas under the curves will always be equal to 1. Probabilities for $\overline{X}$ values can be found in Table A-2 in the Appendix by calculating

$$Z = \frac{\overline{X} - \mu}{\sigma_{\overline{X}}} = \frac{\overline{X} - \mu}{\sigma / \sqrt{n}}$$

where $\sigma / \sqrt{n}$ is the standard of the means and $\sigma_{\overline{X}}$ is the standard deviation of a sample of *n* items.

Example 1

If the drained weights of canned peaches follow a normal distribution with $\mu = 19.0$ ounces and a standard deviation of $\sigma = 0.2$ ounce, what is the probability of a can selected at random having a drained weight of (a) between 19.1 and 19.2 ounces, (b) between 18.7 and 19.1 ounces, and (c) less than 18.8 ounces?

(a)

$$X_1 = 19.1 \qquad Z_1 = \frac{19.1 - 19.0}{0.2} = \frac{0.1}{0.2} = 0.5 \qquad A_1 \text{ (from Table (A-2)} = 0.3085$$

$$X_2 = 19.2 \qquad Z_2 = \frac{19.2 - 19.0}{0.2} = \frac{0.2}{0.2} = 1.0 \qquad A_2 \text{ (from Table (A-2)} = 0.1587$$

$$P(19.2 < X < 19.2) = A_1 - A_2 = 0.3085 - 0.1587 = 0.1498$$

Therefore, about 15% of the drained weights are between 19.1 and 19.2 ounces. (See Fig. 4-9a.)

(b)

$$X_1 = 18.7 \qquad Z_1 = \frac{18.7 - 19.0}{0.2} = \frac{-0.3}{0.2} = -1.5 \qquad A_1 \text{ (from Table (A-2)} = 0.0668$$

$$X_2 = 19.1 \qquad Z_2 = \frac{19.1 - 19.0}{0.2} = \frac{0.1}{0.2} = 0.5 \qquad A_2 \text{ (from Table (A-2)} = 0.3085$$

$$P(18.7 < X < 19.1) = 1 - (A_1 + A_2) = 1 - 0.3753 = 0.6247$$

Therefore, about 62.5% of the drained weights are between 18.7 and 19.1 ounces. (See Fig. 4-9b.)

(c)

$$X = 18.8 \qquad Z = \frac{18.8 - 19.0}{0.2} = \frac{-0.2}{-0.2} = -1.0 \qquad A = 0.1587$$

$$P(X < 18.8) = A = 0.1587$$

Figure 4-9(a)

Figure 4-9(c)

Figure 4-9(b)

Figure 4-9. Areas under the curve for Example 1 (a)–(c).

Therefore, between 15% and 16% of the cans have contents the drained weights of which are less than 18.8 ounces. (See Fig. 4-9c.)

Example 2

In a packing plant grading Satsuma plums whose weights are normally distributed, 20% are called small, 55% medium, 15% large, and 10% very large. If the mean weight of all Satsuma plums is 4.83 ounces with a standard deviation of 1.20 ounces, what are the lower and upper bounds for the weight of Satsuma plums graded as medium?

As $A_1 = 0.20$, the corresponding value from Table A-2 of Z_1 is -0.84, and for $A_2 = 0.25$ the nearest value of Z_2 from the table is 0.67. Then

$$Z_1 = -0.84 = \frac{X_1 - 4.83}{1.20} \qquad Z_2 = 0.67 = \frac{X_2 - 4.83}{1.20}$$

Then, solving for X_1 and X_2, we get

$$X_1 = 4.83 - (0.84)(1.20) \qquad X_2 = 4.83 + (0.67)(1.20)$$
$$= 4.83 - 1.01 \qquad\qquad = 4.83 + 0.80$$
$$= 3.82 \text{ ounces} \qquad\qquad = 5.63 \text{ ounces}$$

Therefore, all fruits weighing between 3.82 and 5.63 ounces should be graded as medium. (See Fig. 4-10.)

Example 3

Number 10 cans of peaches are supposed to hold 72 ounces. The cannery adjusts the filling machine to fill the cans with, on the average, 72.6 ounces. If the distribution of fill weights is approximately normal with a standard deviation of 0.4 ounce, (a) how many of 100,000 cans will contain less than 72 ounces, and (b) assuming that the standard deviation remains unchanged, what mean value should the cannery use for the machine setting if it wants no more than 300 out of the 100,000 cans to contain less than 72 ounces?

Figures 4-11 and 4-12 show that if the filling machine is set to deliver 73.1 ounces, no more than 300 cans per 100,000 produced will contain less than 72 ounces.

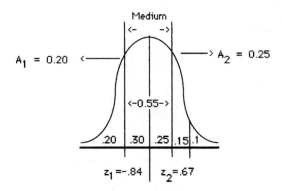

Figure 4-10.   Distribution   chart,   showing medium grade area for Example 2.

$$\mu = 72.6, \quad \sigma = 0.4, \quad X = 72$$

$$Z = \frac{X - \mu}{\sigma} = \frac{72 - 72.6}{0.4}$$

$$= -1.5, \quad P = 0.0668$$

X=72

μ=72.6

$$\text{Number of cans} = 100{,}000(\ 0.0668)$$

$$= 6{,}680$$

Figure 4-11. Example 3(a) calculation and curve.

$$P = \frac{300}{100{,}000} = 0.003, \quad z = -2.75$$

$$-2.75 = \frac{72 - \mu}{0.4}$$

$$\mu = 72 + 2.75(0.4) = 73.1 \text{ ounces}$$

X=72   μ = ?

z=-2.75

Figure 4-12. Example 3(B) calculation and curve.

Example 4

If a lot of 1000 fried chicken dinners has a mean weight of 12 ounces and a standard deviation of 0.6 ounce, (a) what is the probability that, in a random sample of 100 of those dinners, the average weight will be less than 11.90 ounces, and (b) how large a sample must be taken to be 95% sure that the sample mean does not fall below 11.95 ounces?

(a)

$$\mu = 12 \qquad \overline{X} = 11.90 \qquad \sigma = 0.6 \qquad \sigma_{\overline{X}} = 0.6/\sqrt{100} = 0.06$$

$$Z = \frac{\left(\overline{X} - \mu\right)}{\sigma/\sqrt{n}} = \frac{11.90 - 12.00}{0.06} = -1.67$$

From Table A-2, $P = 0.0475$ (see Fig. 4-13).

(b) To be 95% sure that $\overline{X}$ is not less than 11.95 means that the area to the left of $\overline{X} = 11.95$ is $A = 0.05$. From Table A-2, when $A = 0.05$, $Z = 1.645$. Because $11.95 < 12.00$, $Z = -1.645$.

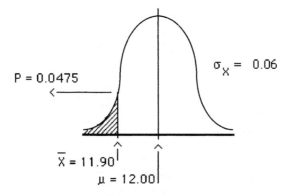

Figure 4-13.   Example 4(a) curve.

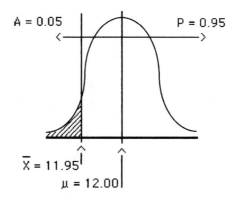

Figure 4-14.   Example 4(b) curve.

$$-1.645 = \frac{11.95 - 12.0}{0.6/\sqrt{n}} = \frac{-0.05\sqrt{n}}{0.6}$$

$$\frac{-0.05}{\sqrt{n}} = -1.645(0.6) = -0.9870$$

$$-\sqrt{n} = \frac{0.9870}{0.05} = 19.74 \qquad n = (19.74)^2 = 389.7$$

Therefore, it will require a sample size of 390 to be 95% sure that the sample mean will not be less than 11.95 ounces. (See Fig. 4-14.)

Normal Approximation to the Binomial Distribution

If n is large and p is small, so that $\mu = np \leq 5$, the Poisson distribution gives a good approximation of the probabilities of the binomial distribution. If n is large and $\mu = np > 5$, the Poisson approximation is less satisfactory, and the normal curve may be used to find such probabilities. The closer together p and q and the larger the value of n, the better the approximation.

The binomial distribution is discrete and its graphical representation consists of adjacent blocks, the areas of which represent the corresponding probabilities. Because the normal curve is continuous, in order to use it to approximate binomial probabilities, the area under the curve must include the block of the histogram centered by any value of r (the number of occurrences under consideration). To include the block centered at r, the value of X to be used in the normal curve equation for the normal deviate must be adjusted by adding $\frac{1}{2}$ to or subtracting $\frac{1}{2}$ from the value of r. The procedure is illustrated in the following example.

Example
If 8% of a packaged product is known to be underweight, what is the probability that a random sample of 100 packages will contain (a) 14 or more underweight packages, (b) 4 or fewer underweight packages, and (c) 5 or more underweight packages?

$$\mu = np = 100(0.08) = 8.0 \quad \text{and} \quad \sigma = \sqrt{npq} = \sqrt{100(0.08)(0.92)} = 2.71$$

(a)

$$r = 14 \quad X = r - \tfrac{1}{2} = 13.5$$

$$Z = \frac{13.5 - 8.0}{2.71} = \frac{5.5}{2.71} = 2.03$$

From the normal curve, $P = 0.212$. Therefore, there is a little better than a 2% chance of this happening. (See Fig. 4-15.)

(b)

$$r = 4 \quad X = r + \tfrac{1}{2} = 4.5$$

$$Z = \frac{4.5 - 8.0}{2.71} = \frac{-3.5}{2.71} = -1.29$$

$$\longrightarrow P = 0.0212$$

$$r = 14$$
$$X = 13.5$$

Figure 4-15.   Example (a) curve.

$P = 0.0985$. There is about a 10% chance of this happening. (See Fig. 4-16.)

(c)

$$r = 5 \qquad X = r - \tfrac{1}{2} = 4.5$$

Because $Z = -1.29$ from part (b) and $P = 1 - 0.0985 = 0.9015$, there is about a 90% chance of this happening. (See Fig. 4-17.)

t-DISTRIBUTION

In all of the preceding examples using the normal distribution, the standard deviation has been known. More often, the variability of a population is not known, and in order to determine the standard deviation, it is necessary to take a sample from the population and estimate the standard deviation σ. Under these condi-

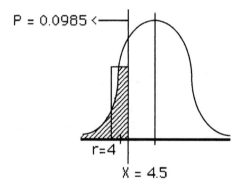

Figure 4-16.   Example (b) curve.

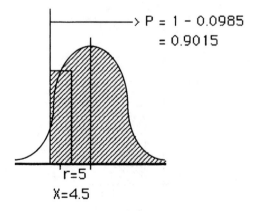

Figure 4-17.   Example (c) curve.

tions, the normal distribution is no longer applicable, and the t-distribution is more appropriate.

The sample mean $\overline{X}$ provides a satisfactory estimate of the population mean μ, but the estimated standard deviation s is estimated using the expression

$$s = \sqrt{\frac{\Sigma(X - \overline{X})^2}{n-1}} = \sqrt{\frac{\Sigma X^2 - \frac{(\Sigma X)^2}{n}}{n-1}}$$

where $\overline{X}$ is the mean of the sample and n is the number of items in the sample.

The quantity $n - 1$ is called the number of degrees of freedom. It represents the number of independent relationships existing among the $X - \overline{X}$ values. Because $\Sigma(X - \overline{X}) = 0$, there are only $n - 1$ independent differences.

Curves representing t-distributions are symmetrical and bell shaped (but not normal), and somewhat flatter—with greater dispersion—than those of the normal distribution. Their shapes depend on the number of degrees of freedom and approach the normal as n becomes increasingly larger.

Values of t depend on the number of degrees of freedom and are shown in Table A-3 in the Appendix for various probabilities. Note that the t-values in the table for infinite degrees of freedom (df $= \alpha$) are identical with the Z-values of the normal curve, and the entries in any column are obviously approaching the corresponding Z-values. From a practical standpoint, the dividing line between a small sample and a large sample is a sample size of 30. with a sample size 30 or over, the normal curve table or t table may be used; below 30, the t table should be used.

Formulas similar to those for the normal distribution are obtained for the t-distribution by replacing σ by s and z by t; that is,

$$t = \frac{\overline{X} - \mu}{s_{\overline{X}}} = \frac{\overline{X} - \mu}{\frac{s}{\sqrt{n}}} \quad \text{(with } n - 1 \text{ df)}$$

satisfies a t-distribution with $n - 1$ degrees of freedom where $\overline{X}$ is the sample mean, and as in the normal curve procedure,

$$s_{\overline{X}} = \frac{s}{\sqrt{n}}$$

Confidence Limits for the Population Mean

The t-distribution is convenient for use in estimating limits for the population mean. If we solve the following equation for μ,

$$\pm t = \frac{\overline{X} - \mu}{\dfrac{s}{\sqrt{n}}}$$

we find that

$$\mu = \overline{X} \pm \frac{ts}{\sqrt{n}}$$

which provides lower and upper limits for the population mean μ with a degree of confidence depending on the value selected for t. These are referred to as the confidence limits for μ.

To find the 95% confidence limits for μ, turn to Table A-3 in the Appendix and select the value of t from the column headed 0.05 with $n - 1$ degrees of freedom (often denoted by $t_{n-1,\,0.05}$). Similarly, 99% confidence limits would be found by using the t-value found in the column headed 0.01 (i.e., $t_{n-1,\,0.01}$).

Example 1
A cannery is supposed to be filling cans with 19.0 ounces of fruit. To check conformance of a new machine with this standard, a random sample of five cans is selected from the new machine's production, and its mean is calculated. If the drained weights of the sample are 18.6, 18.4, 19.2, 18.3, and 19.0 ounces, can one be 95% sure that the production standard is being maintained on this machine? (This is equivalent to asking whether 19.0 is between the 95% confidence limits found for μ.)

X	$X - \overline{X}$	$(X - \overline{X})^2$
18.6	−0.1	0.01
18.4	−0.3	0.09
19.2	0.5	0.25
18.3	−0.4	0.16
19.0	0.3	0.09
93.5	0	0.60

$$\overline{X} = \frac{93.5}{5} = 18.7$$

$$s = \sqrt{\frac{0.60}{4}} = \sqrt{0.15}$$

$$s_{\overline{X}} = \frac{s}{\sqrt{n}} = \sqrt{\frac{0.15}{\sqrt{5}}} = \sqrt{0.03} = 0.1732$$

$$\pm t_{4,\,0.05} = \frac{18.7 - \mu}{0.1732}$$

From Table A-3, $t_{4,\,0.05}$ = 2.776. Therefore, μ = 18.7 ± (2.776)(0.1732) = 18.7 ± 0.48 = 18.22 and 19.18. Because 19.0 is between these two 95% confidence limits, the machine is conforming. (See Fig. 4-18.)

Example 2

The drained weights (in ounces) of a random sample of nine cans of cherries are as shown. Find the 95% confidence limits for the mean weight of the entire lot.

X	$X-\bar{X}$	$(X-\bar{X})^2$
12.1	0	0
11.9	−0.2	0.04
12.4	0.3	0.09
12.3	0.2	0.04
11.9	−0.2	0.04
12.1	0	0
12.4	0.3	0.09
12.1	0	0
11.7	−0.4	0.16
108.9	0	0.46

$$\bar{X} = \frac{108.9}{9} = 12.1$$

$$s = \sqrt{\frac{0.46}{8}} = 0.24 \qquad s_{\bar{X}} = \frac{s}{\sqrt{n}} = \frac{0.24}{\sqrt{9}} = 0.08$$

The 95% confidence limits for μ are $\bar{X} \pm t_{8,\,0.05}\, s_{\bar{X}}$.

$$12.1 \pm (2.306)(0.08) = 12.1 \pm 0.18 = 11.92 \text{ and } 12.28$$

Therefore, one can be 95% certain that 11.92 ounces to 12.28 ounces will contain the mean of the lot from which the sample came.

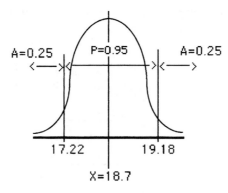

Figure 4-18.   Example 1 curve.

Statistical Hypotheses—Testing Hypotheses

Frequently, we are required to make a decision about populations based on data obtained from samples. For example, we may be interested in knowing if one lot is better than another, or if one production line manufactures better material than another. In order to reach such a decision, it may be useful to start with guesses or assumptions about the two populations. Such an assumption is known as a *statistical hypothesis.* A statistical procedure or decision rule which leads to establishing the truth or falseness of a hypothesis is called a *statistical test.* Decision rules enable the investigator to attach to his decisions probability statements about possible outcomes if the experiment were to be repeated many times under the same conditions. Decision rules are also referred to as *tests of significance,* or *tests of hypotheses.*

The hypothesis to be tested, often denoted by H_0, is called the *null hypothesis,* because it implies that there is no real difference between the true value of the population parameter and its hypothesized value from the sample. For instance, if we wish to determine whether one process is better than another, we would formulate the hypothesis that there is no difference between the two processes, and that whatever differences were observed were merely due to fluctuations in sampling from the same population. Any hypothesis which differs from H_0 is called an *alternative hypothesis* and is denoted by H_1.

In Figure 4-19, the yield for process A is designated by the shaded portion labeled A. After a change is made in the process, the results are plotted in the portion of the chart labeled B. The null hypothesis H_0 states that if the B process falls within the shaded area, then it is no different from process A. The alternative hypothesis H_1 states that if the process should fall outside of the shaded area, then the two processes are different. This very simplistic example has limited usefulness because it does not consider probabilities, but it should illustrate the difference between the two types of hypotheses.

To test the null hypothesis, one assumes that it is true and examines the conse-

Figure 4-19. Null hypothesis for process A and process B.

quences of this assumption in terms of a sampling distribution which depends on it. If we should find that the results obtained from a random sample are vastly different from those expected under the hypothesis, we would conclude that the observed differences were significant and we would reject the hypothesis. For example, if 100 tosses of a coin produced 83 tails, we would be likely to reject the hypothesis that the coin is fair.

The *level of significance* of a statistical test defines the probability level α which is the critical value in decision-making. It is the dividing line between accepting and rejecting. In the event that the calculated probability is less than α, the hypothesis is considered false, and the result is termed *significant*. It is customary to consider a result significant if the probability is less than $\alpha = 0.05$, and to term it *highly significant* if the calculated probability is less than 0.01.

An *alternative hypothesis* is any hypothesis other than H_0 and is denoted by H_1. If H_1 is adopted as the alternative hypothesis, then rejection of H_0 may be regarded as a decision to accept H_1.

If the areas of only one tail of a curve is used in testing a statistical hypothesis, it is called a *one-tailed test*. Similarly, if the areas of both tails are used, the test is referred to as *two-tailed*. Instead of calculating probabilities, one can compare the calculated values of Z with those shown in Figures 4-20 and 4-21. A calculated value which is greater than those shown indicates rejection of the null hypothesis.

Figure 4-20. Normal distribution, 5% level of significance.

Figure 4-21. Normal distribution, 1% level of significance.

The values for Z may be selected from the table of normal curve areas. The critical values for Z most frequently used are for probabilities of 10%, 5%, and 1%. These critical values for Z for both one- and two-tailed tests are tabulated below:

α (level of significance)	0.10	0.05	0.01
Z for one-tailed test	± 1.28	± 1.645	± 2.33
Z for two-tailed test	± 1.645	± 1.96	± 2.58

In testing statistical hypotheses, there is no absolute certainty that the conclusion reached will be correct. At the 5% level, we are willing to be wrong once in 20 times; at the 1% level, once in 100 times. If we are not willing to chance a wrong conclusion this often, we must select an even smaller value for α. Two types of incorrect conclusions are possible. If the hypothesis is true but the sample selected concludes that it is false, we say that a *type 1 error* has been committed. The probability of committing a type 1 error is the relative frequency with which we reject a correct hypothesis, and this is precisely equal to the significance level α.

If it happens that the hypothesis being tested is actually false, and if from the sample we reach the conclusion that it is true, we say that a *type 2 error* has been committed. The probability of committing a type 2 error is usually denoted by β. For a given number of observations it can be shown that if α is given, β can be determined, and that if α is decreased, β is increased, and vice versa. If we wish to decrease the chances of both types of error at the same time, it can be accomplished only by increasing the sample size. α is also called the *producer's* or *seller's risk;* β is termed the *consumer's* or *buyer's risk.* (See Fig. 4-22).

Hypotheses and Decisions Reached from Sample

Decision from sample	H_0 true	H_0 false
Reject H_0	Type 1 error	Correct
Accept H_0	Correct	Type 2 error

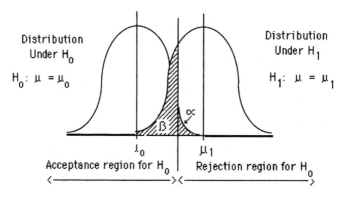

Figure 4-22.   Relation of α and β

Example 1

A company produces frozen shrimp in packages labeled "Contents 12 ounces." A sample of four packages selected at random yields the following weights: 12.2, 11.6, 11.8, and 11.6 ounces. At the 5% level, is the mean of the sample significantly different from the label claim of 12 ounces?

Null hypothesis	Alternative hypothesis
H_0: $\mu = 12$ ounces	H_1: $\mu \neq 12$ ounces

X	$X - \bar{X}$	$(X - \bar{X})^2$
12.2	0.4	0.16
11.6	−0.2	0.04
11.8	0	0
11.6	−0.2	0.04
47.2	0	0.24

$$s = \sqrt{\frac{\Sigma(X - \bar{X})^2}{n-1}} = \sqrt{\frac{0.24}{3}} = \sqrt{0.08}$$

$$s_{\bar{X}} = \frac{s}{\sqrt{n}} = \frac{\sqrt{0.08}}{\sqrt{4}} = \sqrt{0.02} = 0.14$$

$$t_{0.05}(3 \text{ df}) = t_{3,005} = 3.182$$

$\bar{X} = 47.2/4 = 11.8$ ounces
$\bar{X} \pm t_{3,0.05}s_{\bar{X}} = 11.8 \pm (3.182)(0.14)$
$\qquad = 11.8 \pm 0.45$

or

$\mu \pm t_{3,0.05}s_{\bar{X}} = 12.0 \pm 0.45$

Therefore, there is no significant difference between $\bar{X}$ and μ.

Example 2

A new machine in a cannery is supposed to be filling cans of pears with a standard weight of 19.0 ounces. To check conformance with the standard, a sample of 5 cans is selected at random from the production line and its mean is calculated. The drained weights of the samples are shown in the following. Is the sample average significantly different from the standard at the 5% level?

X	$X - \bar{X}$	$(X - \bar{X})^2$
18.6	−0.1	0.01
18.4	−0.3	0.09
19.1	0.4	0.16
18.9	0.2	0.04
18.5	−0.2	0.04
93.5	0	0.34

$\mu = 19.0$ Two-tailed
$H_0 : \bar{X} = \mu$ $H_1 : \bar{X} \neq \mu$
(Null) (Alternative)

$$s^2 = \frac{\Sigma(X - \bar{X})^2}{n-1} = \frac{0.34}{4} = 0.085$$

$$s_{\bar{X}} = \frac{s}{\sqrt{n}} = \sqrt{\frac{s^2}{n}} = \sqrt{\frac{0.085}{5}} = 0.13$$

$\bar{X} = 93.5/5 = 18.7$

From Table A-3, $T_{0.05}$ (4 df) $= t_{4,\,0.05} = 2.776$

95% Confidence limits for μ: $\bar{X} \pm t_{4,\,0.05}\, s_{\bar{X}} = 18.7 \pm 2.776(0.13) = \begin{matrix} 18.34 \\ 19.06 \end{matrix}$

18.34 $\bar{X} = 18.7$ 19.06

The sample is not significantly different from the standard. There are two other possible solutions to this problem:

(a) Find the region within which $\bar{X}$ is not significantly different from μ at the 5% level of significance.

$$\pm t = \frac{\bar{X} - \mu}{s_{\bar{X}}}$$

$$X = \mu \pm t_{0.05}\, s_{\bar{X}} \qquad t_{4,\,0.05} = 2.776$$

$$= 19.0 \pm 2.776(0.13)$$

$$= 19.0 \pm 0.36 = 18.64 \text{ and } 19.36$$

$\bar{X} = 18.7$

18.64 $\mu = 19.0$ 19.36

Region within which $\bar{X}$ is not
significantly different from μ

(b) From the sample mean, find the region within which the population may be expected to fall 95% of the time.

$$\mu = \bar{X} \pm t_{0.05}\, s_{\bar{X}}$$

$$= 18.7 \pm 0.36 = 18.34 \text{ and } 19.06$$

$\mu = 19.0$

18.34 $\bar{X} = 18.7$ 19.06

95% confidence limits for μ

Distribution of the Difference Between Means (Unpaired Observations)

If X and Y are normally and independently distributed populations with means μ_X and μ_Y, and standard deviations σ_X and σ_Y, respectively, then the difference between sample means $\overline{X}$ and $\overline{Y}$ is normally distributed with mean $\mu_X - \mu_Y$ and standard deviation

$$\sigma_{\overline{X}-\overline{Y}} = \sqrt{\sigma_{\overline{X}}^2 + \sigma_{\overline{Y}}^2}$$

where

$$\sigma_{\overline{X}} = \frac{\sigma_X}{\sqrt{n_X}}$$

and

$$\sigma_{\overline{Y}} = \frac{\sigma_Y}{\sqrt{n_Y}}$$

then

$$Z = \frac{\overline{X} - \overline{Y} - \left(\mu_X - \mu_Y\right)}{\sigma_{\overline{X}-\overline{Y}}}$$

will satisfy the normal probability curve.

Usually, σ_X and σ_y are unknown and must be estimated from the two samples. Because their estimates s_X and s_Y will in general not be the same, and because we assume that the sample comes from populations having equal standard deviations, a pooled estimate s is used in the calculations. The pooled estimate is calculated as follows:

$$s = \sqrt{\frac{\Sigma\left(X - \overline{X}\right)^2 + \Sigma\left(Y - \overline{Y}\right)^2}{n_X + n_Y - 2}} = \sqrt{\frac{\Sigma X^2 - \dfrac{\left(\Sigma X\right)^2}{n_X} + \Sigma Y^2 - \dfrac{\left(\Sigma Y\right)^2}{n_Y}}{n_X + n_Y - 2}}$$

then the *t*-distribution is appropriate where

$$t = \frac{\overline{X} - \overline{Y} - \left(\mu_X - \mu_Y\right)}{s_{\overline{X}-\overline{Y}}}$$

with $n_X + n_Y - 2$ degrees of freedom where

$$s_{\bar{X}-\bar{Y}} = \sqrt{s_{\bar{X}}^2 + s_{\bar{Y}}^2}$$

$$s_{\bar{X}} = \frac{s}{\sqrt{n_X}}$$

and

$$s_{\bar{Y}} = \frac{s}{\sqrt{n_Y}}$$

To test the calculated value of t for significance at some level α, it is compared with the tabular value t_α with $(n_X + n_Y - 2)$ degrees of freedom (from Table A-3 in the Appendix). If the calculated value exceeds the tabular value, the means are said to be significantly different at the α level of significance. The usual null hypothesis is $H_0: \mu_X = \mu_Y$, resulting in the tests

$$z = \frac{\bar{X} - \bar{Y}}{\sigma_{\bar{X}-\bar{Y}}} \quad (\sigma \text{ known}) \quad \text{and} \quad t = \frac{\bar{X} - \bar{Y}}{s_{\bar{X}-\bar{Y}}} \quad (\sigma \text{ estimated by } s)$$

Example 1
To determine the difference in butter consumption between rural and urban consumers, data were obtained from 12 farm communities and 14 metropolitan areas. Determine if there is a significant difference in the per capita consumption of butter between the two groups. Because the standard deviation must be estimated from the samples, the t-distribution is appropriate. The hypotheses are $H_0: \mu_X = \mu_Y$ and $H_1: \mu_X \neq \mu_Y$ (two-tailed).

POUNDS		
RURAL X	URBAN Y	
12.1	8.3	
6.8	9.3	
9.1	9.2	
11.1	11.1	
11.4	10.7	
13.3	4.6	
9.8	9.9	
11.3	7.9	
9.4	9.8	
10.2	7.9	
11.3	8.5	
9.8	9.1	
___	9.7	
125.6	6.2	
	___	
	122.2	

$\sum X = 125.6 \qquad\qquad \sum Y = 122.1$

$n_X = 12 \qquad\qquad n_Y = 14$

$\bar{X} = 10.47 \qquad\qquad \bar{Y} = 8.72$

$\sum X^2 = 1345.38 \qquad \sum Y^2 = 1103.37$

$(\sum X)^2 = 15{,}775.36 \qquad (\sum Y)^2 = 14{,}908.41$

$$\frac{(\sum X)^2}{n_X} = 1314.61 \qquad \frac{(\sum Y)^2}{n_Y} = 1064.89$$

$$s^2 = \frac{1345.38 - 1314.61 + 1103.37 - 1064.89}{12 + 14 - 2} = 2.89$$

$$s_{\bar{X}-\bar{Y}} = \sqrt{\frac{2.89}{12} + \frac{2.89}{14}} \quad \text{and} \quad t = \frac{10.47 - 8.72}{0.67}$$

$$= 0.67 \qquad\qquad\qquad\qquad = 2.61 \ (24 \ df)$$

Tabular *t*-values (two-tailed from Table A-3 in the Appendix) are

$$t_{0.05(24 \ df)} = 2.064 \quad \text{and} \quad t_{0.01(24 \ df)} = 2.797$$

Because $t = 2.61$ is larger than 2.064 but less than 2.797, there is a significant difference in butter consumption between the two groups of consumers at the 5% level, but not at the 1% level.

If the question had been, "Is the per capita consumption of the rural group greater than that of the urban group?," the hypothesis would have been $H_0: \mu_X = \mu_Y$ and $H_1: \mu_X > \mu_Y$, and a one-tailed test would have been appropriate. Tabular *t*-values (one-tailed from Table A-3 in the Appendix) are

$$t_{0.05(24 \ df)} = 1.711 \quad \text{and} \quad t_{0.01 \ (24 \ df)} = 2.492$$

Because $t = 2.61$ is larger than both 1.711 and 2.492, the per capita consumption of the rural group is greater than that of the urban group at both levels of significance.

Example 2

Find 95% confidence limits for the mean difference between the performance of two filling machines. Random samples from the two machines, A and B, gave the following results:

$$s^2 = \frac{\Sigma\left(X - \bar{X}\right)^2 + \Sigma\left(Y - \bar{Y}\right)^2}{n_X + n_Y - 2}$$

$$s^2 = \frac{900 + 832}{22} = \frac{1732}{22} = 78.7$$

A	B
$\bar{X} = 350$ grams	$\bar{Y} = 325$ grams
$n_X = 10$	$n_Y = 14$
$\Sigma(X - \bar{X})^2 = 900$	$\Sigma(Y - \bar{Y})^2 = 832$

then

$$s_{\bar{X}-\bar{Y}} = \sqrt{\frac{s^2}{n_X} + \frac{s^2}{n_Y}} = \sqrt{\frac{78.7}{10} + \frac{78.7}{14}} = 3.67$$

95% confidence limits for

$$\mu_X - \mu_Y = (\overline{X} - \overline{Y}) \pm (t_{22,\,0.05}) s_{\overline{X}-\overline{Y}}$$

$$= 25 \pm (2.074)(3.67)$$

$$= 25 \pm 7.61$$

$$= 17.39 \text{ and } 32.61$$

	− 7.61		+ 7.61	
17.39		$\overline{X} - \overline{Y} = 25$		32.61

Therefore, we are 95% confident that 17 and 33 grams are limits of the two confidence intervals that contain the difference in mean filling weights of the two machines. We then compare our specification to see if adjustments are necessary. If $s_X = 9.5$ and $s_Y = 7.7$ had been given instead of $\Sigma(X - \overline{X})^2$ and $\Sigma(Y - \overline{Y})^2$, then

$$s_{\overline{X}-\overline{Y}} = \sqrt{\frac{s_X^2}{n_X} + \frac{s_Y^2}{n_Y}} \quad \text{with } n_X + n_Y - 2 \text{ degrees of freedom}$$

$$= \sqrt{9.02 + 4.24} = 3.64$$

Paired Observations

Many experiments are performed in such a way that each item of one sample is logically paired with a particular item of a second sample. For example, if members of a panel of tasters score a sample of each of two similar products, a set of paired scores results. With n paired scores of X and Y, the differences $D = X - Y$ are analyzed.

The means difference $\overline{D} = (\Sigma D)/n = \overline{X} - \overline{Y}$. If the samples are the same, $\overline{D} = 0$. The question is: Is the average difference $\overline{D}$ too large for the samples to be considered as having the same average scores? To answer this question, one calculates

$$t = \frac{\overline{D}}{s_{\overline{d}}}$$

with $n - 1$ degrees of freedom, where

$$s_{\overline{d}} = \frac{s}{\sqrt{n}}$$

and

$$s = \sqrt{\frac{\Sigma\left(D - \bar{D}\right)^2}{n - 1}} = \sqrt{\frac{\Sigma D^2 - \dfrac{\left(\Sigma D\right)^2}{n}}{n - 1}}$$

The significance of the result is obtained by comparing the calculated value of *t* with the value from Table A-3 (in the Appendix) with the appropriate degrees of freedom.

Example
Using a 12-point intensity scale, ten trained judges scored the bitterness intensity of two samples of beer as shown below. Do the beers differ significantly in bitterness at the 5% and the 1% levels?

$$H_0: \mu_d = 0 \qquad H_1: \mu_d \neq 0 \text{ (two-tailed)}$$

Judge	Beer X	Beer Y	$D = X - Y$	$D - \bar{D}$	$(D - \bar{D})^2$
1	10	7	3	1	1
2	6	3	3	1	1
3	5	6	−1	−3	9
4	7	7	0	−2	4
5	10	7	3	1	1
6	6	4	2	0	0
7	7	5	2	0	0
8	8	6	2	0	0
9	6	3	3	1	1
10	5	2	3	1	1
Totals	70	50	20	0	18

$$\bar{D} = 2.0$$

$$s^2 = \frac{18}{9} = 2$$

$$s_{\bar{d}} = \frac{s}{\sqrt{n}} = \frac{1.414}{\sqrt{10}} = 0.447$$

$$t = \frac{2.0}{0.447} = 4.47$$

$$t_{0.05, \, 9} = 2.263 \qquad t_{0.01, \, 9} = 3.250$$

Therefore, there is a significant difference at both levels.

If we had asked whether beer X was significantly more bitter than beer Y, a one-tailed test would have been appropriate, with

$$H_0: \mu_d = 0 \quad \text{and} \quad H_1: \mu_d > 0$$

where

$$\mu_d = \mu_X - \mu_Y$$

From Table A-3 in the Appendix,

$$t_{0.05, 9} = 1.883$$

and

$$t_{0.01, 9} = 2.821$$

Because 3.16 is greater than both of these values, Beer X is significantly more bitter than Beer Y at both levels of significance.

F-DISTRIBUTION

In applying the t-test for the comparison of two means, a pooled estimate may be made of the population standard deviation on the assumption that $s_X = s_Y$. This assumption usually poses no problems. However, there are cases where the equality of the sample standard deviations is questionable. In such cases, the F-distribution provides a test of equality.

To test the hypothesis that two variances (or standard deviations) are equal, a form of the F-distribution is defined as

$$F = \frac{\text{larger variance}}{\text{smaller variance}} = \frac{s_X^2}{s_Y^2}$$

where

$$s_X^2 = \frac{\Sigma\left(X - \overline{X}\right)^2}{n_X - 1}$$

$$s_Y^2 = \frac{\Sigma\left(Y - \overline{Y}\right)^2}{n_Y - 1}$$

and

$$s_X^2 > s_Y^2$$

In this form, we can compare values of F with critical F-values given in the F-distribution tables. If the calculated value is less than the tabular value, the variances may be considered homogeneous (equal). If the calculated value exceeds the tabular value, the numerator variance is considered to be significantly greater than the denominator variance at the indicated α level. The tabular values of F depend on the degrees of freedom of both the numerator and the denominator. The F-distribution plays an important part in the analysis of variance.

ANALYSIS OF VARIANCE

One-Way Classification

So far, we have found that the t-test provides a procedure for testing the equality of two population means when the data are composed of a random sample from each population. In many cases, more than two population means are under study and they require a method of testing their equality from an evaluation of an independent random sample from each population. For example, we may wish to evaluate the performance of different can closing machine heads, the cost of running a process with several different raw materials, or determining the shipping quality of cases from several different suppliers.

It would appear that these problems could be easily solved by taking all of the possible pairs of samples and testing them individually by the t-test for significant differences between means. This could turn out to be laborious and, worse still, would have a high probability of leading to false conclusions. In a case involving 7 samples taken from populations of identical means, if we tested all the possible pairs of samples for significant differences among their means, we would have to apply the test to 21 pairs, with a 66% chance of arriving at 1 or more incorrect conclusions. We have already discussed the α risk of making a type I error in each individual t-test.

A more efficient procedure for making such comparisons is the *analysis of variance* (ANOVA), which examines all of the sample means together and has a single α risk. Basically, it is a simple arithmetical method of sorting out the components of variation in a given set of data and providing tests of significance. The two principles involved are:

- Partitioning of the sums of the squares.
- Estimating the variance of the population by different methods and comparing these estimates.

Partitioning the Sum of Squares—Equal Sample Sizes

Consider k samples, all of which have n variates (measurements). The notation is summarized in Table 4-5 and introduces double subscripts. The first subscript identifies the sample to which the variate belongs, and the second subscript iden-

Table 4-5.   One Criterion of Classification of Samples

	1	2	3	k
	X_{11}	X_{21}	X_{31}	...	X_{k1}
	X_{12}	X_{22}	X_{32}	...	X_{k2}
	X_{13}	X_{23}	X_{33}	...	X_{k3}
	.	.	.		.
	.	.	.		.
	X_{1n}	X_{2n}	X_{3n}		X_{kn}
Totals	T_1	T_2	T_3	T_k
				Grand total $= T$
Means	$\bar{X}_1$	$\bar{X}_2$	$\bar{X}_3$	$\bar{X}_k$
				Grand mean $= \bar{X}$

tifies the particular variate within the sample. Thus, X_{21} represents the first variate in the second sample, and X_{ij} represents the jth variate in the ith sample.

1. The sums of squares of deviations from the respective means and the corresponding degrees of freedom are defined as follows:

$$\text{SS} = \sum_{i=1}^{k} \sum_{j=1}^{n} \left(X_{ij} - \bar{X} \right)^2 = \sum_{i=1}^{k} \sum_{j=1}^{n} X_{ij}^2 - \frac{T^2}{kn} \qquad df = kn - 1$$

and is a measure of the dispersion of all of the variates about the grand mean.

2. The among-sample means sums of squares SST (or sums of squares for samples) follows from the fact that

$$\sigma^2 = n\sigma_{\bar{X}}^2 \quad \text{or} \quad \sigma_{\bar{X}} = \frac{\sigma}{\sqrt{n}}$$

$$\text{SST} = n\sum_{i=1}^{k} \left(X_i - \bar{X} \right)^2 = \sum_{i=1}^{k} \frac{T_i^2}{n} - \frac{T^2}{kn} \qquad df = k - 1$$

3. The within-sample sum of squares SSE (or sum of squares for error) is the pooled sum of squares obtained within each sample, and the pooled value is unaffected by any differences among the means:

$$\text{SSE} = \left(\sum_{j=1}^{n} X_{1j}^2 - \frac{T_1^2}{n} \right) + \left(\sum_{j=1}^{n} X_{2j}^2 - \frac{T_2^2}{n} \right) + \cdots + \left(\sum_{j=1}^{n} X_{kj}^2 - \frac{T_k^2}{n} \right)$$

$$= \sum_{i=1}^{k} \sum_{j=1}^{n} X_{1j}^2 - \sum_{i=1}^{k} \frac{T_i^2}{n} \qquad df = k(n-1)$$

This quantity is known as the *experimental error* and is a measure of the variation that exists among observations on experimental units treated alike.

From the preceding expressions, it is seen that

$$SS = SST + SSE$$

and the same relationships hold for the corresponding degrees of freedom. Because of this relationship, it is customary to obtain SSE as the difference between SS and SST, and the same with the degrees of freedom. These relationships show that we can always partition the total dispersion in such data into two components—one due to the dispersion existing among the sample means and the other due to the dispersion existing within the samples.

From the sums of the squares and degrees of freedom, we can obtain two estimates of variance σ^2. They are

$$s_{\bar{X}}^2 = \frac{SST}{k-1} \quad \text{and} \quad s_e^2 = \frac{SSE}{k(n-1)}$$

The first of these is influenced by differences among means, whereas the second is independent of any differences; therefore, any difference between these two estimates of σ^2 is the result of differences among sample means. The two estimates may be tested for a significant difference by calculating

$$F = \frac{s_{\bar{X}}^2}{s_e^2}$$

If $s_{\bar{X}}^2$ is significant greater than s_e^2, significant differences among means are indicated.

The *correction term* for the set of variates is defined as

$$C = \frac{T^2}{kn}$$

which occurs in the calculations of both SS and SST. The calculations in the analysis of variance are shown in the following example.

Example

Four scores of each of five products are shown in the following table. Are there significant differences among these mean scores?

	Products				
	A	B	C	D	E
	10	8	5	8	7
	8	7	7	7	6
	9	9	6	7	5
	9	8	5	5	7
Totals	36	32	23	27	25
Means	9.00	8.00	5.75	6.75	6.25

$$C = \frac{143^2}{20} = 1022.45$$

$$SS = (10)^2 + (8)^2 + \cdots + (5)^2 + (7)^2 - C = 1065 - 1022.45 = 42.55 \quad (19 \text{ df})$$

$$SST = \frac{(36)^2 + (32)^2 + (23)^2 + (27)^2 + (25)^2}{4} - C = 1050.75 - 1022.45 = 28.30 \quad (4 \text{ df})$$

$$SSE = 42.55 - 28.30 = 14.25$$

The calculations are usually summarized as shown in the following table.

Analysis of Variance

Source of variation	SS	df	Mean square	F	$F_{0.05}$	$F_{0.01}$
Total	42.55	19				
Products	28.30	4	7.075	7.45	3.06	4.89
Error	14.25	15	0.95	F values at 4 and 15 degrees of freedom		

Because the calculated value of $F = 7.45$ is larger than $F_{0.01} = 4.89$, significant differences among product mean scores are indicated at both the 5% and the 1% levels. Significance at the 5% level is usually indicated by one asterisk, and at the 1% level by two asterisks.

Difference Testing

If the calculated F-value is not significant, further testing is not required. If, however, the F-value indicates significant differences among means, we need to know where the differences exist. Two procedures in common use are the *least significant difference* (LSD) test and *Duncan's Multiple Range* test.

Least Significant Difference (LSD). The difference between the means $\overline{X}$ and $\overline{Y}$ of two independent samples of the same size n, from a normal population with variance $v = s^2$, is tested for significance by calculating

$$t = \frac{\overline{X} - \overline{Y}}{s_{\overline{X} - \overline{Y}}} = \frac{\overline{X} - \overline{Y}}{\sqrt{\dfrac{v}{n} + \dfrac{v}{n}}} = \frac{\overline{X} - \overline{Y}}{\sqrt{2\dfrac{v}{n}}}$$

where $v = s^2$. If, in this equation, we replace t by t_α and solve for $\overline{X} - \overline{Y}$, we have

$$\overline{X} - \overline{Y} = t_a \sqrt{2\frac{v}{n}}$$

which is the boundary line between significance and nonsignificance at the α level and is called the least significant difference between means.

$$\text{LSD} = t_\alpha = \sqrt{\frac{2 \text{ error variance}}{n}} = t_\alpha \sqrt{\frac{2 s_e^2}{n}}$$

where n is the sample size and t_α is based on the number of degrees of freedom for error.

Duncan's Multiple Range Test.   This is a relatively new approach to difference testing and provides a series of shortest significant ranges with which to compare differences between means. The shortest significant range R_p for comparing the largest and the smallest of p means, arranged in order of magnitude, is given by $R_p = Q_p s_{\overline{X}}$, where the values of Q_p can be obtained from Table A-7 in the Appendix. The number of degrees of freedom for the error variance and the standard error (standard deviation) of any mean determined from n individual variates is

$$s_{\overline{X}} = \sqrt{\frac{\text{error variance}}{n}} = \sqrt{\frac{2 s_e^2}{n}}$$

Example
Use (a) the LSD and (b) the multiple range tests to establish significant differences among means at the 1% level for the data of the analysis of variance example.
    In that example, the calculated $F = 7.45$ was larger than $F_{0.01} = 4.89$, so further testing is justified. The error variance is 0.95.

(a)

$$t_{0.01,\, 15} = 2.947$$

and

$$LSD = 2.947\sqrt{\frac{2(0.95)}{4}} = 2.03$$

A	B	D	E	C
9.00	8.00	6.75	6.25	5.75

Any two means not underscored by the same line are significantly different, and any two means underscored by the same line are not significantly different. At the 1% level, means A and B are not significantly different; B, D, and E are not significantly different; and D, E, and C are not significantly different.

(b) $v = 0.95$; p is the number of means being compared, Q_p with degrees of freedom for error from Table A-7 in the Appendix (1% level, 15 degrees of freedom); $R_p = Q_p\, s_{\bar{x}}$ is the shortest significant range for p means; and $s_{\bar{x}} = \sqrt{v/n} = \sqrt{0.95/4} = 0.49$.

Shortest significant ranges

p:	2	3	4	5
Q_p:	4.17	4.35	4.46	4.55
R_p:	2.04	2.13	2.19	2.23

A	B	D	E	C
9.00	8.00	6.75	6.25	5.75

Difference of	Used to compare
2.04	A & B, B & D, D & E, and E & C
2.13	A & D, B & E, and D & C
2.19	A & E, and B & C
2.23	A & C

The results of both methods were the same.

If unequal numbers of measurements are involved in the samples, only minor modifications are required in the analysis. The formula for SST is replaced by

$$SST = \frac{T_1^2}{n_1} + \frac{T_2^2}{n_2} + \frac{T_3^2}{n_3} + \cdots + \frac{T_k^2}{n_k} - C$$

The procedure is illustrated by the following example.

A	B	C	D
9	10	10	11
12	7	8	10
16	6	8	7
15	12	12	7
	7		9
			14
52	42	38	58	$T = 190$

$$C = \frac{(190)^2}{19} = 1900$$

$$SS = (9)^2 + (12)^2 + \cdots + (9)^2 + (14)^2 - 1900 = 152$$

$$SST = \frac{(52)^2}{4} + \frac{(42)^2}{5} + \frac{(38)^2}{4} + \frac{(58)^2}{6} - 1900 = 50.5$$

$$SSE = 152 - 50.5 = 101.5$$

Analysis of Variance

Source of variation	SS	df	Mean square	F	$F_{0.05}$
Total	152	18			
Samples	50.5	3	16.83	2.49	3.29
Error	101.5	15	6.77		

Because the calculated F-value is less than the tabular value for 3 and 15 degrees of freedom, no significant difference among the samples is indicated.

Whenever unequal numbers of measurements occur in the samples and the F-value indicates significance, there is no simple way to decide which pairs of means differ significantly. All possible pairs must be tested individually by means of the t-test.

Two Criteria of Classification
(Randomized Complete Block Design)

In the case of one criterion of classification, no allowance is made for the effects which are common to specific variates in all samples. Where such effects exist, a second criterion of classification is present. This occurs, for example, in a taste-testing experiment in which each judge scores each sample. This design permits another measure of variability—that due to the judges. Then, as before, the total sum of the squares can be subdivided with the addition of a term (SSB) for the judges. Then

$$SS + SST + SSB + SSE \quad \text{and} \quad SSB = \frac{B_1^2 + B_2^2 + B_3^2 + \cdots + B_n^2}{k}$$

where the B's are total scores for the n judges. The pattern is shown in Table 4-6.

Computations

(a) Correction term

$$C = \frac{T^2}{kn}$$

(b) Total sum of squares

$$SS = \sum_i \sum_j X_{ij}^2 - C \qquad df = kn - 1$$

Table 4-6.   Two Criteria of Classification of Samples

Judges (blocks)	1	2	3	...	k	Totals
1	X_{11}	X_{21}	X_{31}	...	X_{k1}	B_1
2	X_{12}	X_{22}	X_{32}	...	X_{k2}	B_2
3	X_{13}	X_{23}	X_{33}	...	X_{k3}	B_3
.	.	.	.		.	.
.	.	.	.		.	.
.	.	.	.		.	.
n	X_{1n}	X_{2n}	X_{3n}		X_{kn}	B_n
Totals	T_1	T_2	T_3		T_k	
						Grand total = T
Means	$\bar{X}_1$	$\bar{X}_2$	$\bar{X}_3$		$\bar{X}_k$	
						Grand mean = $\bar{X}$

(c) Sum of squares for samples

$$SST = \frac{\sum\limits_{i=1}^{k} T_i^2}{n} - C \qquad df = k - 1$$

(d) Sum of squares for judges

$$SSB = \frac{\sum\limits_{j=1}^{n} B_j^2}{k} - C \qquad df = n - 1$$

(e) Sum of squares for error

$$SSE = SS - SST - SSB \qquad df = (kn - 1) - (k - 1) - (n - 1)$$

Example
To determine which of three Napa Valley Chenin Blanc white table wines had the best quality, five expert wine judges evaluated each wine using a 20-point evaluation scale. Did the three wines differ in quality? Did the five judges agree in their evaluations?

	Wines			
Judges	X	Y	Z	Totals
A	16	17	15	48
B	15	16	14	45
C	15	16	15	46
D	16	18	16	50
E	16	16	16	48
Totals	78	83	76	237

$$C = \frac{(237)^2}{15} = 3744.60$$

$$SS = (16)^2 + (15)^2 + \cdots + (16)^2 + (16)^2 - C$$

$$= 3757 - 3744.60 = 12.40 \quad (14 \text{ df})$$

$$SST = \frac{(78)^2 + (83)^2 + (76)^2}{5} - C$$

$$= 3749.80 - 3744.60 = 5.20 \quad (2 \text{ df})$$

$$SSB = \frac{(48)^2 + (45)^2 + (46)^2 + (50)^2 + (48)^2}{3} - C$$

$$= 3749.67 - 3744.60 = 5.07 \quad (4 \text{ df})$$

$$SSE = 12.40 - 5.20 - 5.07$$

$$= 2.13$$

Analysis of Variance

Sources	SS	df	MS	F	$F_{0.05}$	$F_{0.01}$
Total	12.40	14				
Wines	5.20	2	2.60	9.77**	4.46	8.65
Judges	5.07	4	1.268	4.77*	3.84	7.01
Error	2.13	8	0.266			

*Significant at 5%.
**Significant at 1%.

Because the F-values for both wines and judges are significant, we can proceed to determine where differences exist. Suppose that we use the LSD procedure with the 5% level of significance.

Wines:
$$LSD = t_{0.05,\,8}\sqrt{\frac{2v}{5}} = 2.306\sqrt{\frac{2(0.266)}{5}} = 2.306(0.326) = 0.75$$

	Y	X	Z
Means	16.6	15.6	15.2

Therefore, wine Y is, in the opinion of the judges, significantly better than the other two wines.

Judges:
$$LSD = t_{0.05,\,8}\sqrt{\frac{2v}{3}} = 2.306\sqrt{\frac{2(0.266)}{3}} = 2.306(0.421) = 0.97$$

	D	A	E	C	B
Means	16.7	16.0	16.0	15.3	15.0

There is no significant difference between the judging of D, A and E. There is no significant difference between the mean scores of judges A, E, and C, and none between judges C and B.

The analysis of variance procedure is applicable to many different and complicated experimental designs.

5   Sampling

SAMPLING PLANS

In order to measure the quality characteristics of a population—be it a process, a product, or a lot—it is usually desirable to select a sample from the population and examine it. Rarely is it desirable (or even possible) to examine 100% of a lot to determine its quality; therefore, some assumptions must be made so that a rational sample can be selected which represents the quality of the underlying population. Needless to say, if the sample selected does not represent the underlying population, the results of examining that sample are meaningless and can produce incorrect and perhaps costly conclusions. If a sample of cherries is selected from trees on the southern side of an orchard, a study of the degree of maturity of the sample may result in the erroneous conclusion that the orchard is ready for harvesting. The conclusion should have been that only the southern side of the orchard was mature.

The study of sampling is enormously complex, yet analysis of possible sampling procedures ultimately leads to only four simply defined parameters:

1. Where is the optimum location of the sampling point?
2. How should the sample be taken from the population?
3. With what frequency should the sample be selected?
4. What is the optimum sample size?

Methods for analyzing the samples once the system has been established and decisions to be made regarding the population quality based on the results of these analyses are discussed in the chapters that follow. First, let us consider the factors inherent in populations that must be carefully considered so that a reliable sampling plan can be created.

The cherry orchard example illustrates the need to ensure that the sample represents the population and not merely a portion of it. Unfortunately, variables differ from industry to industry and product to product. For example, selecting a completely representative and random sample of corn from the first truckload in the morning and analyzing it for the sugar–starch ratio may provide a plant operations manager with the information required to determine whether to pack whole kernel or cream style—but only for that truckload. Experience might show that the ratio could change drastically in a very few hours; subsequent truckloads might no longer be suitable for the process initially selected. This leads us to the first decision.

WHY SAMPLE?

In the case of the corn, sampling is needed to establish the sugar–starch ratio of the incoming raw material so that the process can be fixed. It is essential that a sample be selected in such a way as to ensure that it represents some defined population. If experience has shown that the ratio changes slowly during the morning for product from a given acreage, it may be possible to sample from the first, third, and fifth trucks from that acreage to establish the ratio, and to observe process drift for the conditions existing on that particular day (wind velocity, temperature, and humidity). Thus, a sample can be used to reliably measure this one corn characteristic for this morning period. In this example the answer to the question *Why sample?* is obvious: It is not possible to determine the ratio of the entire truckload, so a few ears are selected at random in such a manner as to represent the entire population's composition. The difficulties inherent in this example are apparent because of the wide variations possible from farm to farm, from row to row, and even from ear to ear on a single plant. Yet, it is unlikely that a controlled quality canned or frozen corn product could be produced without this critical measurement of the raw material. Many years ago, the tests for suitability conducted in small plants may have consisted of an experienced foreman's selecting several ears from each truckload, cutting a few kernels and chewing them to rate the color, texture, and flavor. Although these tests were inexact and primitive, the conclusions drawn from the sample were as dependent on the relation of the sample to the lot as the conclusions drawn from today's more sophisticated testing methods are dependent on the need for a representative sample.

Sampling a tank truck of liquid sugar or corn syrup for concentration is a far simpler task. It may be shown, for example, that a composite created from 500cc samples drawn from the top, middle, and bottom of a tank truck constitutes a consistently representative sample of all contents. The mere fact that it is possible to obtain a reliable sample with relative ease, however, is not a valid reason to set up a sampling system. The test results should have some meaningful function in the overall product quality system. If, for example, the tank truck arrives with a certified Brix, and if the process requires that every batch of product using that syrup must be adjusted to some unique concentration which is carefully monitored during the process, there may be no advantage to knowing that each shipment is at the concentration expected. This certainly does not preclude the necessity for periodic monitoring, but large volumes of data that are not required to control the quality are of no use and merely add to the cost of quality operations.

The universal definition of sampling is a *procedure used to draw inferences about a parent population from results obtained in the sample.* To the uninformed, sampling is generally considered a simple procedure to discover defects in products, and it does indeed fit that definition on occasion. However, in the examples above, the cherry samples, the corn samples, and the sugar samples were drawn to measure some aspect of the quality of the population from which they were selected; they were not selected to find defects.

SAMPLES FROM DIFFERENT DISTRIBUTIONS

Randomly selecting samples from a population to determine facts about its composition can produce meaningless data if the distribution of data within the entire population is unknown or if the purpose for collecting the data is not clear to the technician. For example, if the heights of a sample of 100 tomato plants are measured, and there are 10 deformed plants only 3 inches high, along with 90 plants that vary from 2 feet to 4 feet in height, an uninformed technician might report the average height of the tomato plants in the field at 2.7 feet. (See Fig. 5-1).

Representative samples taken from any distribution can also be expected to exhibit characteristics of that distribution. For example, representative samples from a bimodal distribution should also have bimodal measurements. Let us consider deliveries of shipping cases received simultaneously from two different plants. The cases have different average bursting strengths and are piled together in a plant warehouse. If truly random samples are selected, bursting strength tests of the samples should show two peaks. (See Fig. 5-2.)

Similarly, samples taken from skewed distributions can be expected to exhibit skewed measurements, mirroring the parent population. It should be noted, however, that compositing samples before analysis, or averaging the results of groups

A portion of the 100 tomato plant sample, showing the effect of including deformed plants in the average height calculation.

Figure 5-1.   Average height of tomato plants.

Figure 5-2. Distribution of samples from two shipments.

of samples, will tend to create normally distributed sample characteristics, thus masking the true character of the population.

SAMPLE SIZE

There are at least five policies commonly used to determine the size of samples selected, and there is a proper place for each of them. Each policy is intended to satisfy two goals: (1) the sample should accurately describe the population and (2) the sample size should be the most economical possible to reach this level of accuracy.

Sample Size of Zero

There are occasions where a sample need not be drawn at all. Some materials are so incidental to the process that there is no need to examine them. Silicones used to lubricate conveyor rails might be an example; or there may be uncomplicated raw material received from a reliable supplier, which will immediately be converted to some other form before use and may not constitute a particularly critical step in the process (i.e., common salt converted to various brines for grading green peas). Other classes of materials used in quantity but requiring no sampling are industrial cleaning solutions and some maintenance supplies. As part of a just-in-time production system, the goal of zero raw material inventories is usually accomplished by shifting the quality control back to the supplier. Examination of the supplier's quality control records eliminates the need for sampling except on a monitoring basis. By far, the greatest use of zero sample size quality control is found in operations where a company depends on luck, supplier reputation, or ignorance. The policy statement, "We have never had trouble with this before, so we don't need to sample it," makes as much sense as the remark made by the farmer whose horse suddenly dropped dead: "That's strange. This never happened before."

So far we have considered only raw materials. Sampling along the production line is a step which cannot be avoided if costs are to be minimized. Yet, there are managers unable to see the value of avoiding the costs of crap and rework by spending a far smaller amount of money on controlling process quality to avoid making mistakes. It is difficult to imagine a situation where it would not be more expensive to correct a faulty finished product than it would be to avoid making the mistake of using an incorrect raw material or preventing errors made on the production line. A zero sampling program of finished product is unthinkable. Yet it makes more economic sense to reduce the sampling at the end of the line (where mistakes are most costly) and increase the sampling on the production line and on the raw materials (where errors are less costly). It would seem almost ax-iomatic that a high production record is nowhere as economically important to a company as a high *acceptable quality* production record.

100% Sampling

At the other end of the spectrum is a sample size of 100%. In Chapter 2, we learned that 100% sampling is not effective. Repetitive sampling is boring and leads to inattention. To demonstrate this point, the author counted all of the *e*'s in the preceding paragraph. The first check produced 70. Rechecks counted 75, 71, and 87. It is suggested that the reader take on the role of inspector and attempt to decide which figure (if any) is correct.

In many instances, 100% inspection is considered necessary because of the importance of some attribute to the consumer. Packages of multiple units can be the source of innumerable consumer complaints about missing product unless some method is devised to perform reliable 100% inspection for a correct count. Tea bags are produced at rates varying from 30 to more than 400 bags per minute and are either machine- or hand-packaged into cartons of many different counts (usually 8, 16, 48, and 100). Even at the lower machine speeds, counting individ-ual bags for hand-filling operations is tedious work and makes accurate counting over an 8-hour shift extremely difficult. Tea bag machine manufacturers have de-vised counting mechanisms that have greatly reduced the chances for miscounted packages. The machines may be equipped with automatic spacers, which can be adjusted to nest stacks of 8 or 16 tea bags so the operator need not count each bag. The operator places one 8-count stack in an 8-count package or two 8-count stacks in a 16-count package, thus virtually eliminating the human error factor. Machines might also have signal devices to indicate missing bags in the stacks, thus alerting the operator to a possible miscount. As a matter of interest, the 100-count packages are generally designed for three rows of tea bags. In hand-packaging operations, each row is loaded with 32 bags (from either 4 stacks of 8 bags, or 2 stacks of 16), and the operator distributes a total of 4 individual bags at the end of the rows to complete the count of 100. The 100% inspection in these instances is performed by the machine and the operator together, with the most difficult part left to the machine.

Where the units have a substantial weight, 100% inspection for count is simplified. Consider a case of 50 packages of 4.0-ounce trail-mix bags. By determining the tare weight of the empty case plus 50 empty trail-mix bags, the correct count can be verified by weighing each case on a 25-pound scale and comparing the weight to the combined tare plus 12.5 pounds of product. Two precautions are necessary, however. First, variation in filler weight deliveries must be controlled; otherwise, 49 bags may be overfilled to such an extent as to produce a case with the proper gross weight for 50 bags. Second, a history of the tare weight must be acquired so the line supervisor can determine how often the scale setpoint needs to be adjusted for tare variations.

Many other methods have been devised to 100% inspect for counts, either direct or indirect. Sensing devices include electric eyes to produce an electric pulse each time a package trips the light beam, timing devices that automatically speed up a line to separate products, or scaling mechanisms attached to condensor plates that cause a change in electric output as a fixed number of products pass over a point in the production line. Signals to initiate electronic counting may also be in the form of proximity switches, or contact microswitches. These and similar systems may be tied to counters or dials to verify counts.

Automatic check-weighers are available to 100% inspect the gross weights of filled packages on the production line. They may be equipped with various degrees of sophisticated readouts or mechanical devices that can calculate quality control charts for averages and ranges and that can remove substandard-weight packages from the line. Others can be electrically or pneumatically tied back to the weighing devices so that automatic weight adjustment can be performed by the check-weigher. Once again, precautions must be taken to assure that tare variations do not interfere with the accuracy of the measuring devices.

Among other 100% inspection devices commonly used are various optical scanning devices such as electric eyes or television cameras. Infrared or visible light beams test for adequate fill height in rigid containers and detect cocked or missing package labels. There is no limit to the list of ingenious devices which may be applied to the 100% inspection concept to dramatically improve the reliability of testing compared to that of the human inspector.

It must be emphasized that no mechanical or electrical system for 100% testing can be installed and allowed to operate without periodic monitoring. The laws of variability that these mechanisms are investigating also apply to the devices themselves. Gears wear, mechanisms stick, metals fatigue, tubes burn out, connections loosen, electric currents vary, levels expand and contract with temperature, and pivot points wear. In a method analogous to the one used to determine sampling frequency, a periodic audit system should be developed by which the testing device can be evaluated before it would normally be expected to allow measurements to drift. The audit should consist of some reliable secondary standard that has proven reliability.

Spot Checking

A quality control system using only spot checking is doomed to fail. Similarly, a quality control system set up without any spot checking is on shaky ground. A spot check is an infrequent, nonscheduled, extra examination of a process or product to reassure the inspector that no mistakes are made and to reassure management that the system is working.

Spot checking can also uncover defects in the quality system. In a plant producing beverage syrups, one of the periodic tests was determination of pH. The quality control supervisor made an informal spot check every other day or so, using a hand-held pH meter. After dozens of uneventful spot checks, the supervisor found a major discrepancy between his reading and that of the line technician. A detailed review of the technician's procedures revealed that the sample had been correctly removed from the line and that the steps required to place the sample in the meter and take a reading had also been done according to instructions. The problem was solved when it was discovered that the buffering agent used to calibrate the meter had somehow become contaminated, resulting in incorrect readings. Fortunately, the plant operated under a batch system, and the few affected batches were isolated and adjusted. As a result of this spot check, the procedure for pH determination was revised to include an extra step for affirming the accuracy of the buffer against a secondary standard.

Spot checks should not take the place of quality audits. A quality audit is a rigorous procedure to ensure that all aspects of a quality system are working satisfactorily. A spot check may be conducted by anyone at any time, using any type of examination as an informal test of a fraction of the quality system, and it has no statistical significance.

Constant Percentage, Square Root

One might expect that replacing the haphazard approach of the spot sample with some sort of a systematized sample would be an improvement, and it frequently is. Without a statistical approach, it is unlikely that a rational system can be devised to provide—with a high level of confidence—a sample that describes the quality level of a population. Shortcuts to statistical determination of sample size are convenient and easy to explain, but they are rarely reliable.

One such shortcut is the use of a constant percentage method. There is no common rule for establishing the percentage. A small percentage is selected if the population (lot size) is large, if quality variation is expected to be minimal, if the samples are to be examined by destructive tests, if the testing is time-consuming or costly, or if the population is made up of expensive units. Conversely, a large percentage is selected if the lot is small, quality is probably nonuniform, testing is simple or inexpensive, or the units are not costly. The errors inherent in such a system should be apparent.

Consider the size of sample required to determine the quantity of oxygen in a tank of liquid nitrogen to be used for gas-packaging peanuts.

The sample will be discarded after testing.
The population (cubic feet of gas) is large.
The oxygen is uniformly dispersed through the tank.
Sampling and testing a large number of samples would be costly.

Even though the last condition (expensive units) is not met, certainly all of the other conditions would dictate taking a single sample, or perhaps duplicate samples, of a very small quantity of gas from the tank. A one-cubic-foot sample might typically represent 0.001% of a small tank's contents. Because of the uniformity of the population, the results of testing this small sample would provide a reliable picture of the gas quality level.

The term *small* means something quite different for different products. A 4000-pound truckload of pineapples might seem to satisfy the conditions for small percentage sampling, but 0.001% sampling might amount to only one pineapple per truckload, certainly not enough to determine the total number of moldy pineapples present. If there are 1000 fruits on the truck, a reasonable number to select might be 10. In this case, a *small* sample would be 1%.

There remain two gnawing doubts about the selection of percentage for sampling: (1) Would a 1% sampling method detect moldy fruit if it were present at 2 per 100, or 6 per 100? (2) Should trucks from ranches with a history of lower quality be subjected to a higher-percentage sampling plan than those with a history of higher quality?

Consider the first question. If 2% of the fruit were moldy, there would be 0.2 moldy pineapples in the sample of 10. There would still be less than one whole moldy fruit (0.6) in the sample if the quality level were three times as poor (6 moldy fruits per 100). In fact, if the sample of 10 disclosed 1 moldy pineapple, it would indicate that the truck contained 10% mold, and this might be far too high to be acceptable. Perhaps a sample of 1% is not the correct number.

Intuitively, one might answer the second question affirmatively. On the other hand, a larger sample might not be required for poorer quality because defects would be easier to find. The problems with using an arbitrary sampling plan should now be apparent.

In an effort to improve on the percentage sampling techniques, some people advocate the use of the square root method. Again, this is a simple method, easily explained to management, which requires smaller sample sizes for large lots than does the percentage sampling scheme. Table 5-1 shows the sample sizes required on two production lines running the same product in different sizes. It assumes that samples will be drawn every hour.

Except for convenience, there is little to be said for selecting the square root system. It has no relationship to the number of defects in the lot and no statistical basis for establishing the confidence level of the results. It assumes that a sample size determined solely by the square root of the lot size is equally able to distinguish between lots with 0.25% defectives and lots with 6% defectives. Unfortunately, this is not true. There is no rational expression for the results of testing lots by this method other than to state that there were x defectives in the samples, or

Table 5-1.   Square Root Sampling Plan

Package size	Production rate	Cans per hour	Samples required	
			3% Plan	Square root
1 gallon	12/minute	720	22	27
1 quart	80/minute	4,800	144	69
1 pint	200/minute	12,000	360	109

that the values for x in the samples tested ranged from y to z. It is not possible to state the level of quality using this method except to state that the lot passes or fails some arbitrary values.

Statistical Samples

So far, we have discussed some of the difficulties associated with types of sampling procedures, and we have implied that there is no way to be sure of a population's quality level from analyzing samples obtained by using these procedures. However, by using statistical sampling techniques, it is possible to describe the quality of the population from which the samples were drawn and to define the probabilities associated with the correctness of that evaluation. The techniques used in statistical sampling procedures is described in detail in the discussion of O.C. Curves (operating characteristic curves) later in this chapter. Statistical sampling procedures answer two very difficult quality questions:

1. How large a sample is required to define the quality of a population?
2. Based on the sample characteristics, is the quality of the parent population satisfactory?

HOW TO TAKE SAMPLES

Sampling procedures must be tailored to the nature of the population being evaluated. The concept of a random sample seems clear enough: A random sample is one that is chosen by a process designed to give every item in the population an equal chance of being chosen. If the population has satisfactory and unsatisfactory product scattered uniformly, then *simple random* sampling is effective. If the population is divided into small subgroups, each containing product of uniform quality, then *clustered random* samples are indicated. If the population consists of layers of good quality and layers containing defects, then *stratified random* sampling is required. Establishing the type of population is usually accomplished by observing the process just before the sampling point. If the type is not easily determined by observation, analysis of groups of samples taken throughout the population will readily define it.

Simple Random Sample

To examine the differences between these types of sampling techniques, let us consider a lot of 5000 cans of green peas, from which we wish to draw a sample of 50 cans in order to determine the drained weight. A possible method of creating a simple random sample might be as follows: Number each of 5000 balls with a different integer from 1 to 5000, with each ball representing 1 can of green peas. Place the balls in a drum, mix, and remove 50 of them. Next, number each of the 5000 cans from 1 to 5000. The numbers on the 50 balls would represent a *simple random sample* of the 5000-can lot. This procedure would satisfy the two requirements for a random sample: (a) that every subset of size 50 *can* be selected and (b) that every subset of size 50 *has an equal chance* of being selected. Obviously, this is an impractical procedure, and only serves to illustrate the principle of simple random sampling. A more useful method would be to remove a total of 50 cans from the line, 1 at a time, intermittently, and spread out over the period required to produce a lot of size 5000. Some bias is bound to be introduced by this compromise procedure. The technician might subconsciously avoid ever selecting a sample from the first 25 cans, or he might sample more slowly at the beginning, faster in the middle, and not at all near the end of the 5000-can run.

The mathematical notation commonly used when referring to the simple random sample is as follows:

Population size: N

Sample size: n

Sample subset: $\begin{pmatrix} N \\ n \end{pmatrix}$

A variation of the simple random sample is the *systematic random sample.* This is generally known as a $\frac{1}{k}$ random sample and is obtained by selecting every kth unit from the population, beginning with a unit selected at random from the first k units. Referring back to the 5000-can lot of peas, we again would number them consecutively as produced and separate them into sublots. If a 1% systematic sample were desired, the 50-can sample would have to be selected in such a way that each of 50 sublots was represented. That means that 1—and only 1—can should be removed from each sublot of 100 cans. In order to be a *random* sample, the can selected from the first sublot of 100 should be randomly chosen. Let us assume that can number 7 was chosen from the first sublot. To be a *systematic* plan, can number 7 should then be selected from each of the remaining 49 sublots. In other words, once a starting point is selected, every 100th can shall be selected, thus generating a total sample of 50 cans.

Note this difference between simple and systematic random sampling: In systematic 1% sampling of a 5000-can stream of production, there are only 100 different sets of 50 cans which may be drawn; in random 1% sampling, *any* combination of 50 cans may be drawn. Systematic sampling provides a much more meager selection of the population. Furthermore, if only the center of each subgroup is to be systematically sampled, then there is only 1 possible set of 50 cans which can be selected. On the other hand, a simple random sampling procedure could conceivably miss the first 200 or 300 cans entirely because of the truly random nature of the selection. The systematic random sample guarantees that each 100 sublots shall be represented in the 50-can sample.

The choice between these two sampling techniques is illustrated in Figure 5-3.

For uniform batch production, use simple random sampling; for continuous production, use systematic random sampling. As with any generality such as this, there are bound to be exceptions. Selection of a system should be tailored to the characteristics of a particular production line. If, for example, cans are filled from six heads, Figure 5-3 would suggest that only head number 3 will be sampled. This may or may not be important to the analysis of the subsequent examination of the samples. If head delivery is to be evaluated, the system should be modified to include each head: select head number 1 for the first sample, head number 2 from the next subset, head number 3 from the next subset, and so on. Alternatively, all 6 heads might be sampled for every 800 cans produced—or some similar system might be randomly selected.

In the preceding example, if one of the six heads were defective, the systematic random sampling technique in which each head is examined will quickly detect the problem. On the other hand, it is possible that a simple random sample system might never select a can from the defective head. Even so, the systematic random sample system is far from perfect. If the samples are taken starting with every 800th can, and a defect due to a malfunction in the feed hopper causes the peas to stop flowing momentarily in between the regular sampling cycle, then the systematic random system might never select a can from this period. Before a sampling procedure is established, the possible causes for the process going out of control should be listed and a sampling procedure that will detect them selected.

We have looked at relatively simple illustrations of the two types of random sampling techniques and have seen that each has hidden weaknesses. This does

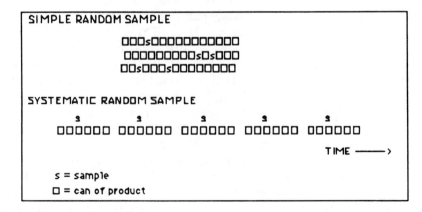

Figure 5-3. Simple random sample and systematic random sample plans.

not mean that the methods are unreliable; rather, it suggests that, no matter which method is used, it should occasionally be audited by use of a procedure that is likely to reveal weaknesses if they exist. One final caution: Because processes are always subject to improvement or accidental changes, the sampling system should be reviewed periodically to verify its effectiveness in assessing the process quality.

Stratified Random Sample

In its classical form, a stratified population is one that is made up of distinct and nonoverlapping groups with differing characteristics. An example might be a hopper (the population) containing a layer (the strata) of mature (the characteristic) blackberries (the group), covered by a layer of slightly reddish immature blackberries, in turn covered by a layer of overripe berries. If samples are drawn at random, a representative sample of the entire hopper would likely determine that the maturity was about average. If samples were drawn from each strata, it is likely that the conclusion would be that the bin contained three maturities. The sampling technique selected would depend on the process which followed the bin storage. If the entire contents were to be mixed into a single batch, a random sample would suffice. If the contents of the bin were to be used continuously, a stratified sample would indicate that the end product would vary as the bin was gradually emptied.

Figures 5-4 and 5-5 illustrate the principle of stratified populations of tomatoes. The pH of each truckload of tomatoes is measured as received at the dock by selecting seven random samples. In this case, each truckload may be considered a strata, and Figure 5-4 clearly shows the variation as received both within each strata and between each strata. In Figure 5-5, the tomatoes contained in five truckloads are considered a single batch, and the 35 random samples show only an approximate range of variation between strata. As with the blackberry bin example, if the truckloads are used in sequence, the pH of the processed tomatoes can be expected to

Figure 5-4.   Stratified samples (seven per truck).

Figure 5-5. Random samples (35 per lot).

fluctuate from 4.2 to 4.1 to 4.3, and so on, as shown in Figure 5-4. The information from Figure 5-5 merely indicates that the tomatoes should average about pH 4.2, and that the process will drift between 4.1 and 4.3. The effect of pH on the quality of the product being manufactured from these raw tomatoes will determine which data (and thus which sampling technique) are required to control the process.

Cluster Random Sample

Cluster random sampling is a technique used when the population is naturally divided into a large number of clusters, where the units within clusters are similar to units between clusters. Under these conditions, a cluster random sample will yield an estimate of the population mean, which is likely to be much more accurate than an estimate based on a simple random sample of the same size.

Consider a production line filling a lot of asparagus into 4800 cans that are then packed in sequence into 200 cases of 24 cans per case. The cases are then stacked in the warehouse. If we wish to take a sample of size 50 (1 can per 96 produced), we are faced with a number of decisions. Are the major quality differences due to variation from can to can, from case to case, from quarter hour to quarter hour, or some other period of production?

To establish the overall quality of the lot, 50 cans could be randomly selected. This is far easier said than done. The cases are relatively inaccessible in their stacks, and they are probably sealed and would have to be destroyed to obtain the samples. Samples removed would have to be replaced with cans bearing the same code numbers, and all 12 cans in each case opened would then have to be repacked and sealed in a new case, which would have to be properly coded and/or stenciled. After all this effort, the sample would represent the average quality of the lot and would not indicate possible variation between cases or variations occurring over the time period produced.

If the cases are considered *clusters,* a group of random cluster samples could be drawn. All of the problems mentioned above would still be present, but they might be reduced somewhat. There are many possibilities for arriving at a sample size of 50, depending on how many clusters are selected, as is illustrated in Table 5-2.

Table 5-2.   Examples of Choices for Cluster Sample Selection

Cluster size	Subsample size	Total sample size
50 cases	1 can from each	50
25 cases	2 cans from each	50
10 cases	5 cans from each	50
5 cases	10 cans from each	50

The variation between clusters is best shown by the use of a 50-case cluster size. The variation within clusters becomes more apparent as the subsample is increased. As with all sampling plans, the goal of the sampling procedure must be established before a rational selection can be made.

It should be emphasized that no cluster sampling technique is the same as a simple random sample from the 4800 cans. The cluster technique includes selection of many subsets, which means that many subsets are not included in the sample. It is also important to realize that one sampling plan does not necessarily provide representative samples for more than one variable. If the variable under consideration is the satisfactory glue application of the label, then the sample should consist of one can from each of many cases. If the variable is the performance of the sealing rollers, then a representative sample might be obtained by examining half the cans in each of a few cases.

TYPES OF SAMPLES

One of the few generalities found in quality control systems is that samples fall into two classes: those in which variables are measured, and those in which attributes are counted. Decisions regarding quality are based on the degree to which variables depart from some normal or desired figure, or the extent to which attributes exceed some specified limit.

Variables are product or process characteristics that can be measured. There are thousands of variables affecting the quality of food products. As an example, consider the variables commonly analyzed to establish the quality of wheat flour:

VARIABLE	MEASUREMENT UNITS
Solids	Percent solids
Moisture	Percent water
Cold-water soluble	Percent
Ash	Percent
Added inorganic phosphate	Percent calcium phosphate
Iron	mg Fe per pound
Calcium	mg Ca per pound
Phosphorus	Percent P

Carbon dioxide	Percent carbon dioxide
Fat	Percent fat
Crude fiber	Percent crude fiber
Acidity of fat	mg KOH per 100 grams
Hydrogen ion concentration	pH units
Sugars	Percent sucrose
Protein	Percent protein
Lipids	Percent lipids
Unsaponifiable residue	Percent unsaponifiable
Starch	Percent starch
Chlorine in flour fat	mg per gram of fat
Nitrite	ppm N
Benzoic acid	ppm benzoic acid
Bromates	ppm bromates
Iodates	ppm iodates
Pigment	ppm carotene
Diastatic activity	mg maltose per 10 grams
Alpha-amylase	Falling Number, seconds
Proteolytic activity	Hemoglobin units per gram

This list refers to product variables and applies specifically to wheat flour. In addition to product variables, there are thousands of process variables affecting the quality of the end product. The following list is far from complete and is intended only to illustrate the array of variables that may be selected for process control.

Temperature	Pressure	Density
Specific gravity	Color	Weight
Flow	Texture	Oxygen
Sediment	Solids	Particle Size
Vacuum	Volume	Headspace
Bacterial count	Mold count	Yeast count
Hardness	Solubility	pH
Vitamin content	Brix	Contaminants
Nitrogen	Viscosity	Flavor
Char	Stone weight	Shell weight
Husks	Additives	Humidity
Carbon dioxide	Chlorine	Level

One characteristic common to all of the variables in both lists is that each item can be measured and expressed in increments of inches, milligrams, degrees, pounds per square inch, percent, and so on. Variables can be measured by use of some type of scale that theoretically can be divided into infinite subunits.

When it is not feasible (or possible) to conduct scalar measurements, another type of quality rating must be used. Often, the most practical method of checking such a process or a characteristic of a product is to classify it as acceptable or unacceptable. This is the method of classifying attributes. Attributes, as contrasted with variables, are quality characteristics of either a product or process, which may be tabulated as either present or absent, satisfactory or unsatisfactory, within or outside of limits, or right or wrong. Note that attributes are *tabulated,* whereas variables are *measured.* These are key descriptors.

Tabulating attributes as the basis for quality control is generally less expensive than measuring variables, but attributes require many more observations to obtain the same information as from a small number of measurements of variables. One reason for this is that it is not generally possible to tell how close an item comes to meeting the desired quality level when it is rejected on an attribute basis.

One of the complications of attribute counting is the need to distinguish between defects and defectives. A defect is a characteristic of either a process or a product which is not in conformance with requirements. A defective is a product containing more defects than allowed. In some instances, this may be as few as one.

It may be convenient to reclassify a variable as an attribute. For example, a package of peanuts may be considered a defect if it exceeds 4.075 inches in length. In this case, a go/no-go gage exactly 4.075 inches long would be used to count the package defects in a lot. This type of sampling and testing might be used if the purpose of the test were to assure that all of the peanut packages would fit into dispensing machines without jamming. If, on the other hand, the goal were to control the amount of packaging material used, exact measurements of package length might be required.

TYPES OF SAMPLING PLANS

The most common procedure for determining the acceptability of the quality level of a lot is to select a random sample, examine it for one or more quality characteristics, tally the number of defects found, and then decide on acceptance or rejection of the lot. The sample size may vary from one unit to many, depending on the size of the lot and the probability of acceptance or rejection. Sampling plans for acceptability of lots are discussed at length at the end of this chapter.

When the decision to accept or reject is based on test results from one sample, the plan is referred to as a *single sampling plan.* In some cases, rejecting a lot on the basis of a single sample may raise doubts in the mind of management, the customer, the vendor, or production personnel. In other cases, a single sampling plan requires destructive testing which might be costly. Because of these considerations, it might be desirable to consider multiple sampling procedures which start with smaller size samples.

In *double* or *multiple sampling plans,* a failure of the first sample to contain an acceptable number of defects automatically permits drawing a second sample (double plans); in the event that fails, additional samples may be drawn (multiple plans).

Double sampling plans may be justified where the costs of the tests are high, because a lot can be considered satisfactory on the first small-size sample if it proves to be defect-free. On the other hand, if the purpose of a double sampling plan is to "give the lot another chance," the principle behind sampling plans is misunderstood. When a lot is rejected, the news is often received with little enthusiasm by production personnel. Their first reaction might be to say, "You pulled a bad sample. Take another and let's see how that one comes out." Depending on the circumstances, perhaps the best answer the quality manager can give to such a request is "maybe we should also pull another sample from each of the last three lots which were found to be satisfactory."

TYPES OF INSPECTION

"Sampling Procedures and Tables for Inspection by Attributes" was originally issued as the Military Standard Number 105 and is available as MIL STD 105E. It has been slightly modified and issued as American National Standard ANSI/ASQC Z1.4 (1981). These standards provide guidance for initiating inspection procedures, as well as instructions on switching to tightened or reduced inspection.

The suggestion that flexible procedures be allowed has been the subject of much debate. The logic is fairly clear, but the rules for switching from one procedure to another are arbitrary and open to considerable question. The principle seems to be that if a producer has a history of acceptable quality, lots from that producer should be examined by a less stringent procedure. The standards state that normal inspection will be used at the start of inspection "unless otherwise directed by the responsible authority." Reduced inspection shall be instituted if all of the following conditions are satisfied:

1. The preceding 10 lots have all been on normal inspection and have all been accepted. (The reason for choosing 10, rather than some other number, is not explained.)
2. The total number of defects found in the samples from the 10 lots does not exceed designated values presented in a table of the standard.
3. Production is at a steady rate. (*Steady* is not defined.)
4. Reduced inspection is considered desirable by the responsible authority. (*Responsible authority* is not defined.)

The standards provide rules for switching from reduced inspection back to normal, normal to tightened, and tightened to normal. In spite of their arbitrary nature, these standards are accepted worldwide and have been very effective in the field of quality control.

CLASSES OF DEFECTS

Classification makes many subjects easier to learn. In the case of defect classification, this is certainly true, but there are other advantages. For one thing, the interrelationships between classes of defects and their respective costs to a company dictate where cost reduction efforts are likely to have the greatest effect. A second reason to classify defects is to determine which types cause the most severe problems with quality control, thus dictating which areas need the most attention.

Quality defects are classified into three broad areas: critical, major, and minor. Critical defects are those which are certain to cause failure of a product to function as designed. The most serious critical defects are those which can endanger the health of consumers—contamination by toxic chemicals, salmonella, botulinum, metal or ceramic fragments, and so on. Other defects (which may or may not directly endanger health) may be classified as critical if they cause the product to become inedible or distasteful—severely discolored, infested with insects, moldy, and so on. A third type of critical defect is one that is in violation of the law and may not be marketed without risk of severe penalties. This classification includes unapproved additives, deceptive labeling, ingredients exceeding regulatory limits, and so on.

The second classification, major defects, is difficult to define. It generally includes defects that are likely to have important adverse effects on the appearance or function of a product. Dented cans might be included in this classification, but the severity of denting could determine the importance of this defect. A dent causing a fracture of the metal is critical; a barely detectable dent might not be considered a defect at all.

A useful criteria for classifying major defects is the probable evaluation that a consumer might be expected to make. If it is likely that a consumer would hesitate to buy a product with a defect which appeared to adversely affect the product quality, that defect is considered major. An example might be a chipped olive jar. One would expect that most consumers would not purchase a chipped jar because of the possibility of glass fragments inside of the container. It might be expected that the consumer would choose an adjacent undamaged jar instead. On the other hand, if one of a group of jars of olives on a store shelf were severely discolored or blackened, the consumer might be expected to select another brand of olives—not another jar of the same brand. The severely discolored jar would then be classified as a critical defect.

The third classification of defects, minor, includes those that would be expected to have only unimportant effects on the performance or shelf life of a product. A slightly scuffed label is a minor defect. Its somewhat undesirable appearance on a package has no effect on the product itself. (A torn or heavily scuffed label, on the other hand, might indicate to a consumer that rough handling of the package could have somehow damaged the product within; this could be a major defect.) Other common minor defects include illegible codes, slightly soiled exteriors, minor blemishes or imperfections, minimal net weight or headspace variation, a few broken or dusty items in a package of friable product, a trace of weeping of a gel, some indication of freezer burn, somewhat stronger or weaker aroma than usual, color a trace lighter or darker than normal, product slightly thick or thin, and so on.

Each industry subclassifies these three classes of defects with language unique to that industry (and often unique to a company within that industry). Over the years, this practice tends to get out of hand; needless subclasses are collected and analyzed for no particularly useful purpose. There are occasions when detailed analysis is valuable, such as the study of a particular defect with the goal of eliminating it. By classifying major defect dents with numerical demerits and tracking them over time and location on the production line, various causes for the defects might be identified, thus leading to elimination of the most troublesome ones.

| | Dent count (500-can sample) | | |
Line location	0–5 mm	6–10 mm	11+ mm
Can cleaner	3	0	0
Filler head 1	2	5	0
Filler head 2	1	1	0
Filler head 3	0	0	17
Filler head 4	2	4	1
Line transfer	5	22	0
Seamer head 1	1	0	0

In this simplified, partial study, it should be apparent that considerable can damage takes place at filler head No. 3 and at the line transfer station. Adjustment or maintenance—or possibly redesign—is indicated.

It has been suggested that converting the three classifications of defects to numerical terms would permit analysis by statistical means. One such scheme is to apply values such as 100, 30, and 10 to critical, major, and minor defects, respectively. Another suggestion is to apply probable scrap or rework costs to each defect as accumulated. It is hard to believe that this type of quality control technique could be useful over a long period of time, but perhaps the use of the dollar analysis for a short-term study might highlight the significance of a problem. It might also provide operations management with the data necessary to justify a capital expenditure to improve the product quality.

Two other important systems for classification of defects are the grading procedure for raw fruits, vegetables, grains, and meats (USDA), and the standards and contaminant regulations for specific foods (FDA). These procedures and standards are rigorously defined and rarely, if ever, require statistical analysis for quality control.

A philosophical note might be in order at this point. There will always be variability in the raw material and the processing of food products. There will always be defects in food products. Some of these defects are preventable; and because quality managers and their corporate leaders are only humans (for the most part), it can be expected that here, too, there will be variability in the selection of projects to eliminate defects. Some projects managers can choose

- Will save more money than others.
- Will not save any money, but will add to costs.
- Might improve the image of the quality department.
- Are very interesting.
- Are extremely difficult.
- Are very important to upper management.
- Might provide solutions detrimental to others' reputations.
- Might apply to short-lived products.

One of the most powerful statistical tools for evaluating priorities is the *Pareto curve,* which suggests that a "critical few" defects produced the bulk of consumer complaints, and that the "trivial many" defects are responsible for a relatively small proportion of complaints. Logically then, the quality department can provide the greatest complaint reduction by eliminating the causes for the "critical few" defects. This is often the case. Bear in mind, however, that among the "trivial many" defects that cause little concern, there are some which could ruin a company if they are not rigorously controlled. One cannot ignore the constant threat of salmonella, for example, even though it is among the "trivial many" defects which rarely, if ever, appears as a complaint.

If the powerful logic of the Pareto analysis does not always work, how does one select the projects necessary to eliminate all defects? The answer is a great deal simpler than most are willing to admit. Yes, Pareto is powerful and useful. And, yes, it is possible to select a project from the preceding list of eight choices. But it is not possible to provide a system that will magically solve all defect problems. The control of quality cannot be established by formula; each company has unique variables of suppliers, processes, equipment, marketing goals, and, most important, its people—with their unique needs, personalities, and motivations. Project selection should be dictated by the needs of the company (not the quality department), and solutions will be created and implemented one at a time. Quality control and quality improvement do not arrive in a bolt of lightning, as so many quality programs promise. Quality control and quality improvement are established and reestablished one step at a time.

SAMPLING RISKS

The risks associated with acceptance sampling have been classified into four groups—two for the producer and two for the consumer. These can be summarized as:

Probability of acceptance
   Producer (Type I risk)
   Consumer (Type II risk)

Probability of rejection
    Producer (alpha risk)
    Consumer (beta risk)

This classification seems at first to be unnecessarily complicated. As we shall see in the discussion of operating characteristic curves (the technique by which an unlimited number of sampling plans can be derived), it is essential that these risks be understood.

Both the producer and the consumer would like to accept only perfect lots: zero defects. Sometimes this is an attainable goal, but only at considerable expense to the producer and at prices which might be unrealistic to the consumer. Each party has to be willing to reach an agreement as to the acceptable number of defective items in a lot. There are no universal rules for determining an acceptable quality level. It is generally arrived at over a long period, during which people from management, production, quality control, purchasing, accounting, research, and sales all contribute ideas until a general consensus is reached. This compromising process is a continuing one. As new needs arise, or as processes, costs, and raw materials change, the acceptable quality level may be either narrowed or expanded.

Having established the acceptable quality level (AQL), the producer and the consumer are now faced with an additional problem: How precise a figure is this? Is there any margin for error? This is where the risk factor appears. Of course, the producer would like to manufacture and ship products containing defects at the AQL level, but there is a risk that the sampling plan will reject some of the lots which are actually satisfactory. This is known as the *producer's alpha risk*. The producer runs the additional risk that the sampling plan will erroneously label lots acceptable that contain a number of defects outside of the AQL. This is referred to as the *producer's Type I risk*. The consumer is faced with similar risks of acceptance and rejection.

SELECTION OF POPULATION TO BE SAMPLED

Determining a sample size is based heavily on statistical concepts. Determining the lot size is equally important, but far more difficult to accomplish. There are no rules and few principles to follow. Let us examine a situation in which a processor receives about 20 truckloads of 100-pound bags of beans per month. Assume that we are interested in assessing the quality level of the following:

Bag weight	Bean size
Insect damage	Foreign material
Mold	Color
Uniformity	Flavor
Moisture	

Truckload Lot

If each truck contains approximately 400 bags, then each truck could be considered a population to be evaluated, and a simple constant sampling procedure could be established. If loads vary from truck to truck, then the population size would require different sampling procedures for each truck. This might unfairly penalize some trucks by requiring more intensive inspection of smaller loads. There are more serious considerations, however. Considering a truckload as a population is satisfactory only if the load is uniform, but one load could consist of produce from several farms or warehouses—with varying quality levels. A sample from such a truckload would average the quality of all bags and might fail to uncover a group of bags with particularly poor quality.

100-Bag (or Other Number) Lot

Arbitrarily dividing each shipment into 100-bag lots will probably reveal the presence of unusual quality portions in a truckload. Such a plan provides average quality information in a continuous stream, thus assisting production personnel in regulating process variables. Variations in quality will probably not be detected by this type of population definition if variations are due to such factors as location of bags in the truck or warehouse before delivery, where the product may have been exposed to heat, moisture, light, or contamination.

Pallet-Load Lot

If there is no particular advantage to isolating portions of truckloads, or if variations between truckloads do not have an important effect on further processing, then loads might be restacked on pallets in the plant warehouse and each pallet load then considered a population. In fact, if the variations are generally unimportant, a population might be defined as two pallets, five pallets, or more. In this case, composite samples can be obtained from several pallets, thus reducing the cost of sampling. Pallets are a convenient population (or subpopulation), but sometimes it might be advisable to dump truckloads directly into bins, in which case the bin contents become the population.

The size of the population selected to establish the quality level of incoming truckloads of beans appears to have no limit. There are, however, a number of guidelines to consider. First and foremost is an understanding of which quality factors must be measured—which factors are important to further steps in the process. If moisture is critical, the first in a series of studies to determine the most logical population size would be a detailed examination of moisture variation within a single bag. If there is no consistent difference between the moisture levels found at the top, bottom, center, and edges of a single bag, then the next step is to examine moisture variation between bags. Similarly, if moisture variation between individual bags taken from a single truckload is minimal, then the third step is to establish moisture level differences between truckloads. This type of investigative testing is continued to establish the points at which significant variations begin to appear. Other possibilities include the study of differences resulting from farm-to-farm, seasonal, year-to-year, and temperature and humidity variations. Once the level at which a significant variation begins to appear is determined, the population size required for quality control sampling for that variable becomes apparent.

Conditions are likely to change; there will always be new trucks, different farms, varietal improvements, different process requirements, and improved packaging of raw materials. Consequently, the procedures for selecting population size must be reviewed periodically to confirm their integrity.

SELECTION OF SAMPLE FREQUENCY AND LOCATION

How often should a sample be selected for quality evaluation? The answer is simple: If the quality variable or attribute to be examined remains within acceptable limits for a period of P units, then the frequency of sampling should be no greater than P units. The value of P should be determined by test, and should be reviewed periodically to ensure that conditions have not changed. Examples of P are the following:

3	truckloads of beans
3	each 250-gallon batches of jam
1	month of warehouse storage
each	supplier
200	jars from sealer
each	label change, at start-up
every	third pallet load

Selecting the location for sampling can be more complicated. Before the advent of statistical quality control methods, attempts to control quality of food production were based on two principles: (1) strict adherence to tried and true procedures and (2) inspection at key points. These procedures have not been totally replaced by statistical quality control. In fact, adherence to procedures is one of the principal tenets of statistical quality control. But mere inspection at key points has been transformed into a far more powerful tool by introducing rational means of evaluating the results of those inspections and tests. Another change has been deemphasizing the importance of finished-case inspection at the end of the line with concentration instead on in-process testing and inspection. Before the advent of statistical quality control, end-of-line inspection worked fairly well: labor was relatively inexpensive, and repair and rework costs were not as critical as they are today. For the most part, companies employing statistical quality control techniques continue to use end-of-line inspection as an audit procedure to provide a scorecard for rating the effectiveness of a quality control system.

HACCP is a food safety system acronym for Hazard Analysis Critical Control Points and has been used by several companies in the United States for over 30 years. The U.S. Department of Agriculture and the Food and Drug Administration have been incorporating this preventative tool into their food regulations. For example, the Code of Federal Regulations for the FDA 21 CFR part 113, which deals with thermally processed low-acid foods in hermetically sealed containers, was first adopted in 1974 and applies the principles of hazard analysis and critical control points. Critical control points are covered in great detail: personnel, equipment, procedures, containers, closures, preparation, processes, thermal operations, deviations of critical control factors, records, and reports. In 1981, the FDA published suggestions for a formalized HACCP program in which they recommended preparing a detailed process flow diagram to help locate key sampling points (process hazards and critical control points.) It is expected that, ultimately, HACCP will be applied universally.

First, a process flow diagram (Fig. 5-6) is constructed for each product, showing the individual steps in sequence from receipt of packaging and raw materials, through plant processing steps, to shipment or storage of the finished product. Then each step is analyzed for possible hazards. The FDA defines a process hazard as "a possible source of trouble along the processing chain which might be defined as a failure to identify: (1) critical materials, (2) critical processing points, (3) adverse environmental conditions, and (4) human malpractices."

Once these critical control points have been determined, a sampling system including sample location, size, frequency, selection methods, testing, analysis, and reporting is established. From management's point of view, inadequate control of critical control points is undesirable (or even fatal), because it exposes the company to the dangers of producing defective products which might, if shipped to the public, adversely affect the company's reputation or even the financial structure. At the very least, defective products could result in costly scrap or rework; at worst, they could result in product recall or adverse consumer action.

In addition to these immediate concerns, there are legal considerations as well. Failure to recognize critical control points might result in unacceptable risks of adulteration—or failure to comply with federal, state, or local regulations regarding net content, food standards, good manufacturing practices, sanitation, or storage conditions.

Figure 5-7 is another example of a HACCP chart. This one is for quality control of empty cans received on pallets, can ends, insert coupons, and plastic resealers. In this application, cases, case glue, coder ink, and other case supplies are handled on a separate HACCP chart (along with other cardboard and flexible packaging supplies). The numbers listed after critical equipment refer to quality control standard procedures as described in the quality control manual. For example, 1.01 refers to laboratory tests for approval of can lithography (color and content), dents, leaks, side seam and throwback solder, leaks, foreign material, scuffs, and so on. For companies with standardized written operating procedures, another set of numbers might also appear on the HACCP chart, referenced to applicable operations to be taken by plant personnel. This set might include cursory physical examination, count and tally with receiving documents, checkoff tickets with loads and incoming paperwork, and similar operations documents.

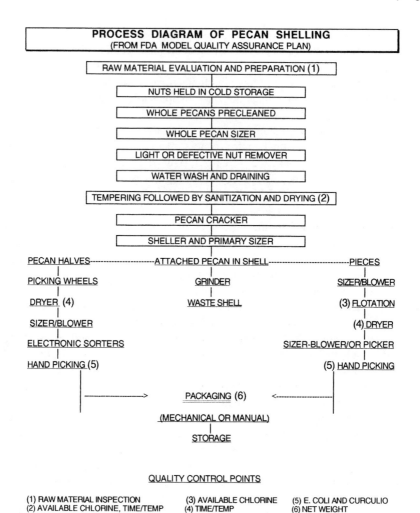

Figure 5-6. Process diagram of pecan shelling (from FDA model quality assurance plan).

In any business endeavor, the management must be aware of the possible hazards in the conduct of the business which will adversely affect the outcome; and they also realize that some of these hazards are critical to satisfactory performance. To this extent, HACCP is not a unique concept, but it differs from informal procedures for avoiding mistakes and failures in that it is a detailed system which incorporates seven distinct areas.

• Identification of severity of a likely hazard.
  The FDA and the USDA state that a hazard may be a biological, chemical, or physical property that may cause a food to be unsafe for consumption.

PROCESS DIAGRAM OF BULK CAN SUPPLIES

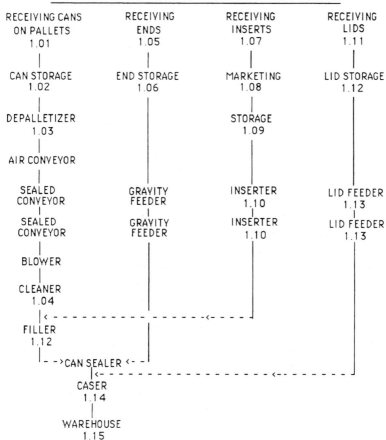

Figure 5-7.   Process diagram of bulk can supplies.

Additionally, a commercial hazard is any condition which would make the product dangerous or unsalable.

- Locating the critical control points (CCP) in which a failure would likely result in a hazard being created or allowed to persist.

    A CCP is a point or procedure in a food system where loss of control may result in an unacceptable product (or in the case of FDA and USDA, an unacceptable health risk). Controls applied at these process steps shall prevent, eliminate, or reduce to acceptable levels the hazards identified.

- Documenting control limits for each point.

    Tolerances of safety shall be established for each CCP to control a quality hazard (defined by FDA and USDA as a health hazard, or multiple health hazards).

- Creating a system for monitoring these critical control points.

    Each CCP shall be controlled through a planned sequence of measurements or observations which are recorded for subsequent audit procedures. Continuous (100%) monitoring is recommended, but where it is not possible, the frequency of observation should be determined by statistically designed data collection systems. A system of calibrating the equipment, reagents, and instruments used in controlling and monitoring should be included. This also applies to validation procedures of software for computer control systems.

- Establishment of corrective actions to be taken.

    When a critical control limit has been exceeded, the cause of the deviation must be corrected and eliminated. The materials involved shall be tagged, or otherwise identified, and removed from the process for subsequent disposition. Each CCP will require specific corrective action in the HACCP plan. Precise records of each deviation and corrective action taken shall be maintained for the shelf life of the product.

- Preparing and maintaining records of these observations.

    A written HACCP plan (manual, procedure) should be prepared, and continually updated. Records generated during operation of the plan should be current and complete. A check sheet consisting of "pass/fail" is not considered satisfactory for recording data at critical control points. Instead, actual values obtained during monitoring should be recorded, along with product description, product code, and time and date of data collection. Consumer complaints relating to CCP should be included in the HACCP records.

- Providing routine audits of the system to verify its adequacy.

    Audit requirements of the government may differ from those of a processor. Audits may include nonscheduled spot check sampling; review of CCP records for completeness; updating the HACCP plan to conform to changes in materials, process, product, packaging, storage, or distribution.

Acceptable control limits for these tests and analyses, along with preventative measures for each control point, should be documented. Corrective action and record keeping are required for an effective system. Finally, a system of audit procedures will ensure that the HACCP system is working satisfactorily.

Figures 5-6 and 5-7 illustrate relatively uncomplicated flow charts. Most processes are far more complex, and require considerable effort for use in designing a HACCP system. One of the most effective methods of creating a HACCP process flow chart is the use of team effort. A preliminary flow chart describing a manufacturing process from raw materials to the feed bin might start out as shown in Figure 5-8.

After a team of personnel from several departments have worked on this simplified flow chart, it might well result in the document shown as Figure 5-9.

Note that Figure 5-9 shows several *control points,* but no *critical control points.* This distinction is important. For example, control point CP2.200 covers all of the raw materials for the product. At the bottom of the list is Color R,Y, with the test CO (or colorimeter) listed in the adjoining block. From the FDA viewpoint, approved color is not a critical control point because it is not a health haz-

Figure 5-8.   Simple flow chart.

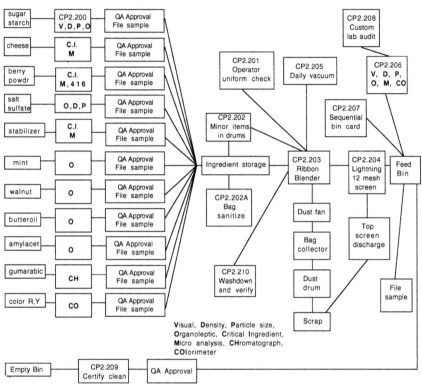

Figure 5-9.   Detailed process flow chart, showing control points.

ard to the consumer. On the other hand, an incorrect, sickly yellow color additive (instead of, for example, a specified deep orange color required for the product) can certainly be considered critical to the acceptance of the product quality. In addition, there are several raw materials which are critical to the manufacturer and require organoleptic tests to assure that the product quality will be acceptable; but these materials are otherwise *safe* to use and would not be considered critical by the FDA or USDA.

There are two solutions to this dilemma. First, for a company with sufficient personnel to handle the extra work involved, two separate systems can be designed. One for the FDA/USDA, in which only critical control points are shown, and another in which all points necessary to maintain quality control are shown. This would also require two separate methods manuals, and two separate data files.

For smaller companies, it would be more efficient to design a single flow chart with *all* of the control points shown. To distinguish the critical control points from the other quality control points, the CCPs might be differentiated by coloring the control point boxes or, as shown in Figure 5-10 by using extra heavy borders on the boxes considered critical. Again, two sets of records need to be kept so that the HACCP records are readily available for audit at any time.

Each HACCP flow chart requires an accompanying written explanation of each of the critical control points. For the chart illustrated in Figures 5-9 and 5-10, a format is suggested which identifies each critical control point, those responsible for testing and approval, the control procedures, and documentation required. There may be several other control points which are not of a critical nature, but it is suggested that these *not* be included in the HACCP manual.

Note that although references are shown for the control procedures, sample size, sample frequency, test methods, specifications, tolerances, and report forms, they are not detailed on this HACCP document (Figure 5-11). Since these techniques tend to be fairly complex and voluminous, including them here would un-

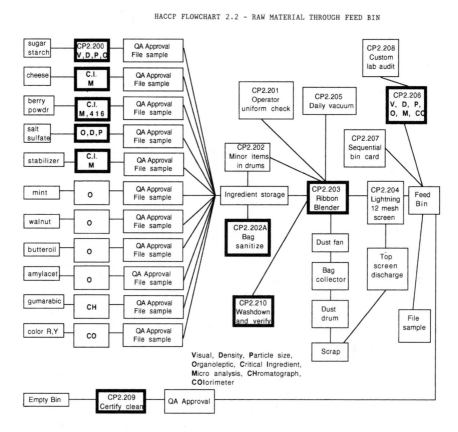

Figure 5-10. Combination flow chart of control and critical control points.

HAZARD ANALYSIS AND CRITICAL CONTROL POINT MANUAL   **Page** xx
CONTROL POINT CODES

PRODUCT FAMILY 2.2
PRODUCT: ROMAN HOLIDAY PUDDING FLAVOR BASE MIX
PROCESS: RAW MATERIAL THROUGH FEED BIN

CP Code	2.200	2.201	2.202
Identification	Raw materials	Personnel	Drums
Prime Responsibility	Quality assurance	Process supervisor	etc.
Secondary Responsibility	Purchasing	Personnel	etc.
Control Procedures*	QA V #10.97 QA D #10.88 QA P #10.60 Taste Test O #6.95 QA #416 R&D M #3.22 QA CO #10.86 R&D CH #3.15	House Rules No. 4, 6, 15, 22 (Uniforms and sanitation)	etc.
Documentation	Lot number form and ticket. Hold/release form	Checklist San#1	etc.

* The numbers refer to standard test procedures in various departments.

Figure 5-11.   Suggested format for HACCP manual.

necessarily complicate the HACCP manual. Furthermore, these procedures should be readily available in manuals prepared by Quality Assurance, Production, Engineering, Purchasing, and other departments.

Of necessity, the above explanation of HACCP programs is rather generalized. It is expected that governmental regulations will become very detailed and increasingly specific as they are refined over the years. For example, the seafood regulations will not be applicable to the bakery industry or the canned foods industry except in a general way. The critical control point problems tend to be peculiar to each industry: stones in raisin processing, grain toxins in the baking industry, cooling-water contamination in aseptic canning plants, proteolytic decomposition in meats, pesticide residues in vegetables, frozen food refrigeration failure, and so on.

The identification of critical control points is a relatively straightforward procedure. In the food industry, each step of the process can be examined and the CCPs identified: raw materials (chemical, physical, or biological fabrication, harvesting, and growing); receiving and storage; production steps (processing and packaging), distribution (including warehousing); and final consumption. Hazard analysis, on the other hand, contains some judgmental characteristics which are more difficult to evaluate. Some of the systems offered for analyzing microbiological hazards, for example, categorize the hazards according to the risk severity. *Clostridium botulinum* is classified as a severe hazard; *Staphylococcus aureus* is classified as a moderate hazard; and yet, neither one can be permitted in a food product.

Perhaps the same problem exists with chemical hazards. All chemicals can be toxic at some concentration, even table salt. Some cannot be permitted at any level in food products, whereas safe limits have been established for others. Safe levels of physical contaminants are even more difficult to evaluate. How much is "an acceptable level of a little dust"? How coarse is dust which is classified as "unacceptable grit"? 21CFR Part 109 attempts to specify definitions and tolerances for unavoidable contaminants in food and food packaging materials. Unfortunately, over the years, these tolerances have been misunderstood by the general public.

A final emphasis: The governmental safety regulations, detailed though they may be, do not cover the critical control points which govern market acceptability of the food product. Net weight control, product color, flavor, texture, aroma, granulation, flowability, solubility, and dozens of other product characteristics may have no bearing on safety but could mean the difference between an acceptable product or a market failure. Any one of these characteristics may be considered a critical control point in the manufacture of a specific product. Most companies have been well aware of these CCPs for years, and as pointed out above, they must continue to be monitored, either separately or as part of the overall HACCP system.

This examination of the many aspects of sampling might seem disproportionately lengthy. However, a quality control program assembled without a thorough understanding of the nature of sampling will accomplish little, if anything. Precise tests followed by accurate statistical analysis and detailed reports may be worthless if they are based on samples which do not represent the population under study.

To obtain a truly representative sample as a basis for establishing a useful quality control system, there are probably only four absolute requirements:

- Determine the location of sampling points critical to the population under consideration.
- Establish a sampling method which will represent the population characteristics.

- Using statistical methods, select the sample size which will produce results with the probability needed.
- Specify the frequency of sampling based on the expected cycle of population quality variation.

ATTRIBUTE SAMPLING PLANS

References have been made above to sample size, operating characteristic curves (O.C. Curves), sampling plans, types of inspections, and sampling risks. These subjects are contained in the military standard, "Sampling Procedures and Tables for Inspection by Attributes," known for years as MIL STD 105E. Recently, it was adopted in slightly revised form by the American National Standards Committee of ANSI, and issued as ANSI/ASQC Z1.4 (with the same title) (1981).

Either of these standards provides an extremely flexible series of sampling plans which encompass 15 lot size classifications, 7 inspection levels, 26 acceptable quality grades, 3 degrees of inspection severity (normal, tightened, reduced), as well as double and multiple sampling. Operating characteristic curves, showing producers and consumers risks for each plan, are also included. In spite of this flexibility, the standard appears to be difficult to use; it is actually quite simple. Once quality and inspection parameters are agreed upon, selecting the specific sampling procedure is easy. The standard has long been internationally accepted for use in purchasing contracts by both seller and buyer. Its most serious drawback is the difficulty of explaining to nontechnical people how the tables are constructed and why they work.

To arrive at a sampling plan, the following steps are taken (see Tables A-9 and A-10 in the Appendix):

1. Decide on the lot size to be examined and select the corresponding code letter from the Code Letter table. Inspection level II is normally used. (For example, lot size 1000 had code letter J.)
2. Find this letter in the single sampling plan master table for normal inspection, and note the corresponding sample size. (For example, code letter G requires a sample size 32).
3. Decide on the acceptable quality level (for this example, let us select 1.5%) and find the accept/reject level. (For letter Q, and 32 samples, with acceptable quality level 1.5%, the lot of 1000 would be accepted if 1 or less defects were found, and would be rejected if 2 or more rejects were found.)

Detailed instructions and definitions are found in the first several pages of the standard, including techniques for special sampling plans.

It is suggested that these special plans be avoided, or used with great care. They were probably introduced to limit the costs of testing when either the test itself was costly or time-consuming, or when destructive testing of several expensive products was prohibitive (e.g., destructive testing of an engine, a parachute, or complex electronic assemblies). For most food products, these are not particularly important considerations. There is a psychological approach to double sampling plans: They erroneously seem to present a "second chance" to the testing of the acceptability of a lot. (If it fails on the first test, draw additional samples.) To the statistically informed, this is nonsense. The multiple sampling plans become quite complex in their usage, and there is a tendency for the uninformed to attempt to use shortcuts, which makes the results meaningless.

With these precautions in mind, the standard provides a useful attribute sampling plan for industry. In the introductory remarks, it suggests several inspection applications: end items, components and raw materials, operations or services, materials in process, supplies in storage, maintenance operations, data or records, and administrative procedures. The user is cautioned to consult the operating characteristic curves contained in the standard to find a plan which yields the desired protection (risk).

6   Test Methods

To cover thoroughly the subject of test methodology in a single chapter would be close to impossible. There are hundreds of books available describing thousands of test methods for myriad food products and their components. Provided here is a brief discussion of some of the relationships between test methods and quality control, along with a bibliography for guidance.

Tests might be classified in three groups, addressing:

- Legal requirements.
- Process, product, packaging specifications.
- Special guides.

Legal requirements exist for nearly all food products. These include net contents, microbiological purity, nutritional claims, and absence or presence of food or nonfood components. Following are a few FDA requirements selected at random from the Code of Federal Regulations:

- If vitamin A is added to milk, each quart must contain not less than 2000 International Units.
- Egg rolls must contain more than 2.56% (by weight) whole egg solids.
- Each part of the contents of a package of margarine must bear the word *margarine* in type or lettering not smaller than 20-point type.
- Ninety-nine percent disodium ethylenediaminetetraacetate may be used up to 500 parts per million to promote color retention in canned strawberry pie filling.
- When processing bottled drinking water, cleaning and sanitizing solutions used by the plant shall be sampled and tested by the plant as often as is necessary to assure adequate performance in the cleaning and sanitizing operations.

In these few examples, there is no question as to what tests should be conducted. Certainly, the first step in selecting tests is to find the legal requirements at all levels of government (some city and state codes exceed the requirements of the federal laws).

When selecting test procedures not legally prescribed for controlling the quality of a process, product, or package, methods may be rigorously standardized or developed in house along scientific (but less stringent) lines. There are also many arbitrary in-house methods, which are not recognized by industry or government but which serve a specific purpose in controlling quality. The most important test of all is the user test. If a product meets all known legal requirements and company specifications but does not work, the company is in serious trouble. Obviously, the specifications are incomplete. A painful example happened in a company who had successfully manufactured tea bags for years. The bag specifications appeared complete—bag dimensions, 35 grains of product per bag, 3-inch string length, ¾-inch square tag size, uniform ¼-inch seam width, seal complete on three sides; and so on. Suddenly an unexpected torrent of consumer complaints poured in, all describing the string tearing loose from the bag. The problem was traced to insufficient wet strength in a shipment of tea bag paper—a variable for which there was no specification. The quality laboratory test was scientifically designed to evaluate all of the dimensions listed above—plus a final product test in the cup. Unfortunately, the scientific test did not include lifting the brewed tea bag from the cup by its string. The problem would have been revealed the day the first roll of defective paper was received if the lab had realistically duplicated the consumer's method of preparing a cup of tea.

A basic test method library should include the books listed below under "Basic Library." Many of these references will not be applicable to all food products, but it might be well to review each for possible modification. Some tests may be adapted to nonspecific products by merely changing the sample preparation. Following the basic library is a list of "Recommended References" commonly used in food laboratories. Next is a group of "Test Methods" references for various analyses. The last list suggests a few "Industry Test Methods." By contacting suppliers, it is possible to obtain other test procedures that have been developed by private industry and by trade associations.

BASIC LIBRARY

American Public Health Association. *Standard Methods for the Examination of Water and Wastewater.*
Food and Drug Administration. *Bacteriological Analytical Manual.*
Food and Drug Administration. *Microscopic Analytical Methods in Food and Drug Control.*
Food and Drug Administration. *Training Manual for Analytical Entomology in the Food Industry.*
Grant, E.L. and Leavenworth. *Statistical Quality Control,* 6th ed., McGraw-Hill.
Gruenwedel, D. and J. Whitaker. *Food Analysis.* Dekker.
Harrigan, W. *Laboratory Methods in Food and Dairy Microbiology.* Academic Press.

Joslyn, M. *Methods in Food Analysis*. Academic Press.

Marriott, N.G. *Principles of Food Sanitation*. Chapman & Hall.

National Research Council. *Food Chemicals Codex*. National Academy Press.

Pomeranz, Y. and C. Meloan. *Food Analysis Theory and Practice*. Van Nostrand Reinhold.

Potter, N.N. and J.H. Hotchkiss. *Food Science*. Chapman & Hall.

Richardson, G. *Standard Methods for Examination of Dairy Products*. American Public Health Association.

Schilling, E.G. *Acceptance Sampling in Quality Control*. Marcel Dekker.

Speck, M. *Compendium of Methods of the Microbiological Examination of Foods*. American Public Health Association.

Williams, S. *Official Methods of Analysis*. Association of Official Analytical Chemists.

Zweig, G. and J. Sherman. *Analytical Methods* (12 Volumes). Academic Press.

RECOMMENDED REFERENCES

General Analysis

Aurand, L. et al. *Food Composition and Analysis*. Van Nostrand Reinhold.

Birch, G. *Analysis of Food Carbohydrate*. Elsevier.

Chaplin, M. and J. Kennedy. *Carbohydrate Analysis: A Practical Approach*. Oxford.

Fennema, O. *Food Chemistry*. Dekker.

Fresenius, W. *Water Analysis*. Berlin.

Gilbert, J. *Analysis of Food Contaminants*. Elsevier.

Kurtz, O. *Micro-Analytical Methods for Food Sanitation Control*. Association of Official Agricultural Chemists.

Macrae, R., R.K. Robinson, and M.J. Sadler. *Encyclopedia of Food Science, Food Technology and Nutrition*. Academic Press.

Pomeranz, Y. and C.E. Meloan. *Food Analysis*. Chapman & Hall.

Special Instrumentation

Gilbert, J. *Applications of Mass Spectrometry in Food Science*. Elsevier.

Hui, Y.H. *Encyclopedia of Food Science and Technology*. Wiley.

Lawrence, J. *Food Constituents and Food Residues: Chromatographic Determination*. Dekker.

MacLeod, A. *Instrumental Methods of Food Analysis*. Halsted.

Microbiology

American Public Health Association. *Recommended Methods for the Microbiological Examination of Foods*.

Beuchat, L. *Food and Beverage Mycology*. Van Nostrand Reinhold.

Corry, J. et al. *Isolation and Identification Methods for Food Poisoning Organisms.* Society for Applied Bacteriology.
Post, F. *Laboratory Manual for Food Microbiology.* Star.
Sharpe, A. *Membrane Filter Food Microbiology.* Wiley.

Sensory

Bourne, M. *Food Texture and Viscosity: Concept and Measurement.* Academic.
Lawless, H.T. and B.P. Klein. *Sensory Science Theory and Applications in Food.* Dekker.
Moskowitz, H. *Food Texture: Instrumental and Sensory Measurement.* Dekker.
O'Mahoney, M. *Sensory Evaluation of Foods: Statistical Methods and Practices.* Elsevier.
Piggott, J. *Sensory Analysis of Foods.* Elsevier.
Stone, H. and J. Seidel. *Sensory Evaluation Practices.* 2nd ed. Academic.

TEST METHODS

Aurand, L. and A. Woods. *Food Composition and Analysis.* Van Nostrand Reinhold.
Banwart, G. *Basic Food Microbiology.* AVI.
Charalambous, G. *Analysis of Food and Beverages.* Academic.
DiLiello, L. *Methods in Food and Dairy Microbiology.* AVI.
Gorman, J. *Principles of Food Analysis for Filth, Decomposition and Foreign Matter.* Food and Drug Administration.
Harrigan, W. and M. McLance. *Laboratory Methods in Food and Dairy Microbiology.* Academic.
Jay, J. *Modern Food Microbiology.* Van Nostrand Reinhold.
Jellinek, G. *Sensory Evaluation of Food.* VCH Pubs.
Jowitt, R. *Physical Properties of Food.* Elsevier.
King, A. *Methods for Mycological Examination of Food.* Plenum.
Knorr, M. *Food Biotechnology.* Dekker.
Lee, F. *Basic Food Chemistry.* AVI.
Macral, R. *HPLC in Food Analysis.* Academic.
Morris, B and M. Clifford. *Immunoassays in Food Analysis.* Elsevier.
Moskowitz, H. *Applied Sensory Analysis of Foods.* Chemical Rubber Co.
Mountney, G. *Practical Food Microbiological Technique.* Van Nostrand Reinhold.
Okus, M. *Physical and Chemical Properties of Food.* American Society of Agricultural Engineering.
Osborne, B. and T. Fearn. *Near IR Spectrographic Food Analysis.* Wiley.
Post, F. *Laboratory for Food Microbiology and Bacteriology.* Star.
Rockland, L. and G. Stewart. *Water Activity.* Academic.
Stewart, K. and J. Whitaker. *Modern Methods of Food Analysis.* AVI.
Vaughn, J. *Food Microscopy.* Academic.

INDUSTRY TEST METHODS

Evaluating a Double Seam. Dewey and Almy Chemical Company (a division of
W. R. Grace Company). Cambridge, MA.

FPA Technical Manual of Specifications and Test Procedures. National Flexible
Packaging Association. Cleveland, OH.

Index of CCTI Recommended Industry Standards and Testing Procedures. The
Composite Can and Tube Institute. Washington, DC.

Official Microbiological Methods of the American Spice Trade Association.
ASTA. Englewood Cliffs, NJ.

Specifications and Methods of Analysis. N. V. Cacaofabriek De Zaan. Koog-
Zaandijk, Holland.

Hundreds of test methods developed specifically for food analysis are available from instrument manufacturers. When contemplating additions to laboratory
equipment, check with the manufacturer's technical department to obtain lists of
applications. Contact the manufacturer of presently owned equipment for new applications developed since the original purchase. Although all of these industry
tests may not be accepted as official analytical methods, they may provide useful
rapid analyses on such equipment as:

Atomic Absorption Spectroscopy—concentration of inorganic elements.

Biosensors—concentration of glutamic acid, sugars, organic acids, cholesterol.

Capillary Electrophoresis—identify amino acids, proteins, nucleic acids, organic
and inorganic ions, vitamins, peptides.

Chemiluminescence—determine presence of nitrogen compounds.

Colorimeter—evaluate color (ripeness, bake, roast, browning).

Computerized Viscosimeter—collects data and plots viscosity versus temperature.

Differential Scanning Colorimeter—protein denaturation, gelation, oil oxidation.

Dynamic Rheometer—viscosity measures at elevated temperatures.

Electrophoresis—separate biological compounds polysaccharides.

Elemental Analysis—identify elements in organic materials.

Flame Atomic Absorption—identify specific atoms.

Fluorescence Spectroscopy—components of an organic material.

Fourier Transform Infrared Spectroscopy—identify organic compounds in a sample; also protein, fat, moisture, amines.

Gas Chromatography—isolate compounds within mixtures, detect additives, pesticides, fatty acids.

Gas Pycnometry—density of particles.

Immunoassay Kits—identify animal species; determine contaminants such as in-
secticides, toxins, pesticides, drugs in dairy products.

Inductively Coupled Plasma Emission Spectrometry—test for inorganic compounds, trace elements

Inductively Coupled Plasma Mass Spectrometry—determine presence of large numbers of elements in single analysis

Infrared Moisture Meter—loss-of-weight moisture analyzer.

Ion Chromatography—analysis of ions such as sulfate, nitrate, bromate.

Ion Selective Sensors—measure pH, chloride, nitrate.

Laser Diffraction—analyze particle size.

Liquid Chromatography—identify vitamins, sugars, aflotoxins, amino acids.

Liquid Chromatography Mass Spectrometry—identify individual compounds in mixtures.

Liquid Spectrophotometer—color measurement of liquids.

Magnetic Resonance—fat content, moisture, freezing processes.

Mass Spectrometry—identify compounds, including molecular structure.

Microwave Extractor—rapid extraction of cholesterol, fatty acids.

Modular Rheometer—shear stress of pastes and viscous liquids.

Near Infrared Analyzer—multicomponent analysis of solids, pastes, slurries, moisture, fat, oil, sugars, protein.

Photon Correlation Spectroscope—measure submicron particle size.

Super Critical Fluid Extractor—fat analysis.

pH Meters—pH, oxidation–reduction potential

Polarimetry—analyze antibiotics, proteins, organic acids, vitamins, sugars.

Rotational Viscometer—measure viscosity and flow.

Supercritical Extractor—rapid extract of fats and oils.

Thermal Analysis—determine moisture content, melting points.

Ultraviolet and Visible Spectrophotometer—analysis of organic compounds.

Viscometers—viscosity.

Water Activity Meter—dew point, vapor pressure, temperature, and water activity.

7　Product Specifications

Assembling specifications to describe a product would seem to be the simplest part of a quality control program. A bread bakery might decide to develop a cake to round out the product line. Once the kind of cake has been decided on, a brainstorming session with the bakery staff can produce a list of attributes which need to be defined. (See Fig. 7-1).

In order to convert product attributes into specifications, three basic questions must be answered:

- What do you want to make?
- What are you capable of making?
- How much can you afford to spend?

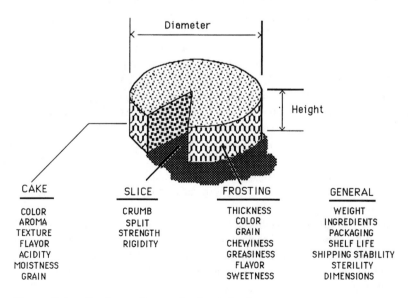

CAKE	SLICE	FROSTING	GENERAL
COLOR	CRUMB	THICKNESS	WEIGHT
AROMA	SPLIT	COLOR	INGREDIENTS
TEXTURE	STRENGTH	GRAIN	PACKAGING
FLAVOR	RIGIDITY	CHEWINESS	SHELF LIFE
ACIDITY		GREASINESS	SHIPPING STABILITY
MOISTNESS		FLAVOR	STERILITY
GRAIN		SWEETNESS	DIMENSIONS

Figure 7-1.　Quality attributes of a frosted cake.

Converting descriptive quality terms into concrete specification numbers can be accomplished logically by the use of statistical tools. A statistical evaluation of process capability will define the limitations of cakes the plant can consider manufacturing. Capability studies will also lead to decisions regarding the numerical values to be applied to each of the variables under consideration: weight, moisture, height, color, acidity, shelf life, and cost. The range of attributes can be determined by formulating experimental designs and analyzing the results: texture, grain, rigidity, sweetness, flavor, and so on. Compromises suggested by the experiments regarding shelf life, packaging, ingredients, dimensions, and product characteristics can then be effected through discussions with marketing, accounting, distribution, engineering, production, and other departments. Finally, the specifications can be presented to management for approval.

Now that the cake and frosting specifications are spelled out, the details of the ingredients, packaging, processing, and storage must also be specified in order to bake and market cakes with the consistent quality level required. Although these specifications usually require fewer department meetings, statistical design and analysis of experiments are again the surest paths to reliable data. If any of the above steps for converting ideas into product specifications is omitted, the disappointment of quality failure on the production floor and in the marketplace becomes likely. If specifications are either overlooked or unrealistic, quality failure on the production floor is bound to cause high scrap and rework costs; quality failure in the marketplace is bound to cause loss of sales.

Listing specification data for an existing product is a simpler task than writing specifications for raw materials or ingredients produced by others. Selecting ingredients from the vast maze of materials available in the marketplace can seem overwhelming. Is it possible to select and then specify an ingredient which has the performance, flavor, uniformity, price, stability, and purity characteristics necessary to provide the specific performance, flavor, uniformity, price, stability, and purity characteristics desired in the finished product? Assuming that such an ingredient exists, is it then possible to assign to it the required specification data which suppliers can meet consistently? The answer to both questions is yes. For the sake of expediency, it may be practical to establish greatly simplified and fairly broad specifications at the outset; through continuous use of a program of statistical experimental design and analysis, specifications can be fine-tuned to assure uniform quality.

The techniques of experimental design and analysis, process capability study, and sensory testing will be discussed later. For the present, let us examine the content of specifications. An ingredient specification for cocoa is presented in Figure 7-2. The numbers shown are illustrative only and represent values which might be required for a frosting mix ingredient.

To arrive at the specification in Figure 7-2, the Boston Biscuit Bakery (BBB) completed several series of tests in which each of the measurements listed was found to be required either for the quality cake desired or to comply with laws and regulations. A great deal of information is available from cocoa manufacturers, but not all available analyses are necessary in the ingredient specification. For example, composition data are available for 14 chemical elements, 7 vitamins, ox-

BOSTON BISCUIT BAKERY
Chocolate Cake #CC071 Ingredient Specification

INGREDIENT: Cocoa powder
        #FCP 071

APPROVED SUPPLIERS:
        Andrus Cocoa Co.   Supl#C3
        Filbert Specialties Supl#C1

DATE APPROVED: Sept. 20, 19--

APPROVED BY: *ACB*

DESCRIPTION: Medium fat cocoa, conforming to the standard of identity as described
in 21CFR 163.112 and 163.113, and produced in accordance with Good Manufacturing
Practices detailed in 21CFR Part 118, Cacao Products. Gras Clearance 21CFR 182.1.

ATTRIBUTE		MEASUREMENT	TEST METHOD
PHYSICAL	Color	B2-B4	Agtron A#76 Phototron P#3
	Flavor	Betw. Std A and Std B	BBB Method #0449
	Fineness	99% minimum	Wet, thru 200 Mesh sieve (0.075mm)
	Odor	Free from off odors	BBB Method #0006
	Shell Content	5% maximum	BBB Method #0720
CHEMICAL:	Fat Content	10.0 - 12.0%	AOAC 13.036
	Lecithin	4.0 - 5.75 %	BBB Method #0724
	Moisture	5.0% maximum	AOAC 13.003
	pH	7.0 - 7.4	AOAC 13.008
	Lipase activity	Negative	BBB Method #0726
MICROBIOLOGICAL:			
	Std Plate Count	5,000 maximum	FDA/APHA
	Molds per g	50 maximum	BBB Method #0728
	Yeasts per g	50 maximum	BBB Method #0729
	Enterobacteriaceae	Negative in 1 g	BBB Method #0730
	E. coli per g	Negative	FDA/APHA
	Salmonellae	Negative	FDA Bacty Manual

PACKAGING:
        50-pound multiwall paper bags, polyethylene-lined. Each bag labeled
        with product name, manufacturer, production lot number.

STORAGE, SHIPPING:
        Shipped with no more than 2 sublots per delivery. Transported according
        to 21CFR 110.80. Pallets to be stored in cool, dry (R.H. less than 50%) area,
        isolated from spices and flavorings. Shall comply with storage
        conditions in accordance with FDA good manufacturing practises.

SAMPLING PROCEDURE:
        Samples shall be taken according to BBB Procedures specified in
        sampling manual #BBB Manual #4.

Figure 7-2.   Cocoa powder specification.

alic acid, theobromine, starch, sugars, sulfates, and organic acids—all of which may be found routinely in cocoa products. There is no point in listing these data in the specification merely because figures are available. Specification requirements should be concerned only with those variables that might affect the quality of the final product.

Similar specifications are prepared for each ingredient in the standardized BBB cake formulation. This ensures that the starting conditions for product manufacture will be correct. This is the second of the three classes of specifications required for a quality control system. In addition to the ingredient specifications and the finished product specification discussed above, the need for a process specification should be obvious.

Process specifications are proprietary and usually in a continuous process of evolution as new machinery, raw ingredients, process tests, and formulations are discovered. In the case of the BBB cake, process controls are evolved for each of the unit operations:

Ingredient weighing
Blending
Mixing
Depositing
Baking
Cooling
Packaging

Process specifications for each operation are developed from process capability studies, and test methods are improvised as needed to assist in measuring the attributes considered critical to the process. As discussed under HACCP, control points are discovered by analyzing the process steps and their effects on the end product. Both the product and the process equipment are included in the process specification network. For example, time–temperature relationships during the baking cycle must be measured and recorded for both the oven equipment and for the cake itself during the baking cycle.

Process specifications are the most complex of the three types. Even starting with raw materials which fall within the ingredient specifications, small variations in the flour composition coupled with trace variations in the leavening agents and minor differences in atmospheric pressure can combine to upset the normal leavening process and thus create an in-process oddity not easily corrected by unbending process controls.

Each industry seems to collect its own group of such oddities. The density of a dehydrated food product was found to shift markedly each afternoon, even though no known changes in the ingredients or the process were occurring. The cause was finally found to be an afternoon wind blowing against the dehydrator air inlet which upset the drying airflow. In a french fry plant, raw potatoes tested satisfactorily when received but turned darker than specifications permitted when processed. It was discovered that stock was not being properly rotated: As a re-

BOSTON BISCUIT BAKERY
Chocolate Cake #CC071 Process Specification

DATE APPROVED: Sept. 20, 19--   APPROVED BY: ACS

INGREDIENT WEIGHING

Small Ingredients   Scale S2007 Precision 0.25 ounce

          Butter flavor B03   3 lbs 7.75 oz
          Albumen powder   74   8.00
          Vanillin   1   4.25
          Salt REF17 S   20   2.00
          Mix 4.5 minutes on Ingredient Blendor #BI02. Temperature 65-80 deg.
          Sodium tartrate ST04   5   0.00
          Calcium phosphate CP04   2   0.25
          Add to Ingredient Blendor #BI02. Blend 1.75 minutes. Maximum 90 deg.
          Store in sealed drums at 65-90 deg. for no more than 8 hours.

     Major Ingredients   etc. etc.

Figure 7-3. Partial process specification.

sult, some of the raw potatoes had started to increase their sugar content in storage—and thus fried darker. Another mystery went unsolved for years in a tea packaging operation: Periodically, packages would appear with unusually high (and undesirable) dust content. All ingredients and process controls were within specification, but the problem persisted. Eventually, it was discovered that dust was accumulating in the ends of a lengthy feed hopper above the scales, then abruptly emptying into the production line packages. This type of exceptional line or product performance makes process specifications difficult to compile.

The section of the process specification illustrated in Figure 7-3 may read more like a recipe than a process control. Subsequent data concerned with airflows, purge times, temperatures, oven speeds—and interactions of these variables with product variables—are often more descriptive. Because the moisture content of the cake is a critical variable, a process specification has to include a table of expected values of moisture levels for various combinations of oven conveyor speed and temperatures in both primary and secondary sections of the oven. Likewise, a process specification covering the cooling cycle needs to list either the cooling time or the cooling conveyor speeds at various ambient temperatures in order to provide uniform product quality.

To summarize:

1. Start with a description of the ideal product.
2. Through studies of available ingredients and process experimentation, define attainable product specifications.
3. Prepare optimum ingredient specifications.
4. Prepare optimum process specifications.

8   Process Capability

Before examining the mechanics of process capability analysis, it would be wise to define the subject. With only two words to define, this would appear to be a trivial task. According to the dictionary, the word *process* refers to steps (or operations, or changes) leading to a particular result, and *capability* is defined as the capacity for an indicated use. Let us examine a few more definitions of process capability, proposed by various authorities in the field.

"Process capability is the measured, inherent reproducibility of production turned out by a process." (Juran, *Quality Control Handbook,* 1988)

"The minimum spread of a specific measurement variation which will include 99.7% of the measurements from a given process." (Hansen, *Quality Control Theory and Applications,* 1963)

"The minimum range of values of the process output variable which includes at least 99% of the observations under normal operating conditions." (Braverman, *Fundamentals of Statistical Quality Control,* 1981)

"Process capability is quality-performance capability of the process with given factors under normal, in-control conditions." (Fiegenbaum, *Total Quality Control, Engineering and Management,* 1983)

"Process Capability—the level of uniformity of product which a process is capable of yielding." (DataMyte Corporation, *DataMyte Handbook,* 1987)

"Process capability is determined by the total variation that comes from common causes—the minimum variation that can be achieved after all special causes have been eliminated." (Deming, *On Some Statistical Aids Towards Economic Production,* 1975)

No two of these definitions are in complete agreement; nor do they specify whether they are concerned with a complete product, a single step in a process, or a vendor's product quality (raw material), or if a machine can be considered to be a process. One definition refers to the "spread of a measurement," another, the "reproducibility of production," a third, "range of values," and "quality performance," "99.7% of the measurements," "minimum variation," and so on. Perhaps the most useful definition is the simplest one—"the level of uniformity of product which a process is capable of yielding." This definition does not concern itself with how well the process is currently working, but rather with what a process is *capable* of yielding. The implication here is that, at times, the process works more uniformly than at other times; more important, there is a possibility that the process might be further improved. Herein lies the value of process capability studies: They will establish the best quality a process is currently capable of producing so that realistic control limits and specifications can be drawn up. That is the immediate goal. The studies required by capability analysis will also suggest areas where modifications might bring further improvement.

Deming refers to causes of two types of variability in manufacturing processes: common and assignable. Common causes are inherent in the process and will remain there until management sees fit to change the process in order to modify or eliminate them. Assignable causes are external factors that can be eliminated or controlled, and although they affect the process, they are not a part of it. Examples of assignable causes of variability are incorrect packaging material, improper identification of raw material, and wrong laboratory data. Examples of common causes include poor lighting, gradual machine wear, inadequately trained operators, and inexact measuring equipment. *Process capability studies are conducted on processes which have all assignable causes eliminated, and insofar as possible, have all common causes reduced to a minimum.* By definition, these conditions exist when a process is in *statistical control.* Herein lies the problem: More often than not, a process capability study is required for new or proposed processes which are not yet free of as-yet-unidentified assignable causes. This might explain the conflicting definitions. Each quality control practitioner has conducted process capability tests under unique conditions for which he later evolved a definition to suit his purpose.

Example

Let us examine a simplified example of a process capability study. The example is taken from a plant processing melba toast. One of the quality criteria is the color development in the final stages of oven browning. Individual pieces of toast are scanned by a spectrophotometer which has been calibrated to reflect yellow-brown light over a reduced spectrum. The readout is expressed in percent of a standard yellow-brown reference plate, and a value of 41 is considered optimal. Data is collected every 30 minutes starting at early morning and continuing for 15 hours. The results are tabulated and graphed in Table 8-1 and Figure 8-1, respectively. Superimposed on the graph are the calculated values for the average and for the upper and lower 3-sigma control limits (AVG, UCL, and LCL). These limits are considered totally unsatisfactory; the upper limit is far too light, and the product tested at the lower limit is nearly burned. There are many variables at work during this 15-hour period which contribute to the unsatisfactory range.

To limit these variables, another set of data is obtained in which samples are taken consecutively, 15 seconds apart, until 30 samples have been obtained. The samples are then examined, data recorded and graphed, and control limits applied (see Fig. 8-2). Now we

Table 8-1. Example of Process Color Capability Data

Sample no.	Color units every 30 minutes	Color units consecutive	Color units improved proc.
1	34	42	41
2	45	44	41
3	62	41	41
4	35	44	42
5	44	45	41
6	33	43	45
7	42	44	42
8	50	44	43
9	44	41	41
10	53	41	40
11	33	38	39
12	32	39	38
13	57	40	39
14	45	41	40
15	40	41	41
16	33	40	41
17	45	43	42
18	20	40	41
19	44	44	42
20	31	38	42
21	59	39	41
22	34	40	40
23	52	44	40
24	29	42	40
25	45	45	41
26	44	42	42
27	52	44	41
28	33	43	41
29	34	47	42
30	29	42	41
	Average 41.1	Average 42.0	Average 41.0
	Standard dev. 10.0	Standard dev. 2.23	Standard dev. 1.29
	3 × Std. dev. 30.1	3 × Std. dev. 6.7	3 × Std. dev. 3.8
	Control limits 11.0–71.2	Control limits 35.3–48.7	Control limits 37.2–44.8

Figure 8-1. Color units every 30 minutes.

Figure 8-2. Color units, consecutive.

find control limits within 6.7 units of the average value of 42.0. Contrast this with the range of 30 units in either direction of the average observed in the 15-hour sampling period. What we have discovered are the capability limits of the process when most variables are unchanging. In other words, the process under the best conditions now available will process toast within a range of 13.4 color units.

During the course of the study thus far, certain observations of the process may have been questioned, which suggest possible ways of reducing variability. For example, it may have been observed that the widest swings in data over the 15-hour test period were the result of the coarse control setting on the trim burners; that is, the sensitivity to temperature control was set at the lowest range, causing the burners to alternately surge and then coast. This resulted in marked differences in product color units. Suppose the range setting on the burner was moved up a notch so that the burners cycled more often as the temperature started to drift. With this improved process control, another set of data can be taken (see the last column in Table 8-1 and Fig. 8-3). Here we find that the consecutive sample series showed reduced variability and improved process capability; the control range was reduced to 7.6 color units.

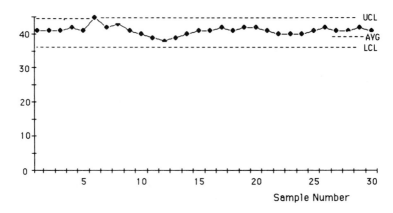

Figure 8-3.   Color units, consecutive (improved process).

This example demonstrates how a process capability study can be made using simple statistical control charts. It also demonstrates how process improvements can be discovered, implemented, and verified. However, it should be emphasized that this example is simplified. In the real world, there are at least three major changes to be considered:

- Thirty samples are not enough. Generally, at least 100 samples will be required to obtain a reasonable estimate of population mean and standard deviation.
- A subgroup size of one is more difficult to calculate than larger sizes. Subgroups of three or five may be used, thus permitting the use of tables to estimate standard deviation. Bear in mind that here we are dealing with means, not individuals, and results should be interpreted with constant reference to the *means*.
- After conducting the first series of determinations to establish process capability, there is rarely an opportunity to run process improvement tests *immediately,* as shown in the example. These require time to plan adequately.

A word of caution regarding the second point: Not all quality control practitioners agree that there is justification for considering the control limits for means as synonymous with process capability. Even so, it may be practical, at least initially, to use the determination of means control limits to explore the capabilities of a process. Eventually, it might be necessary to measure the variability and control limits of individual products; after all, the customer examines the product purchased, not a subgroup of products. In fact, the customer does not care about average quality—only about the quality of an individual product purchased. This is not necessarily true for intermediate products, however. In a corn cannery, for example, there is no need to know ear-to-ear variability; bushel-to-bushel quality data may readily suffice.

A more complete determination of process capability may be arrived at by investigating the time-to-time variability of the process. The consecutive data are perfectly valid, but they define the best possible control limits of the process as it presently operates. These data would be more valuable if some measurement of process shift could also be measured. Figure 8-1 shows no particular drift pattern but clearly indicates that the process does not meet the control limits determined in the consecutive process capability study. Specifically, if the capability control limits of 35.3–48.7 were applied to the every-30-minutes data, 11 points would be out of control on the low side, and 7 points would be out of control on the high side. That means a total of 18 of the 30 points are out of control. This process is obviously in need of attention.

By contrast, the consecutive data show a rather tightly controlled process over the period of time required to obtain the data. Somewhere between that period and 30 minutes, the process drifted several times. To locate that specific interval, a detailed test in which samples are taken at fairly short intervals for a total time of 30 minutes must be conducted. To simplify calculations, it is suggested that the average color units of every five samples (in order of production) be plotted on a graph. A visual examination of the graph ought to reveal the interval during which the process is stable, and the approximate time after which the process drifts in one direction or another. Once this interval has been established, the causes for drift can be studied to reduce variability. Without taking this step to determine the minimum drift interval, attempts at process improvement testing are likely to be disorganized, random, and most likely fruitless.

A graph showing process drift is shown in Figure 8-4. It appears that the shortest interval between a stabilized process and the start of a drift is about 7 minutes. Therefore, the process should be evaluated every 7 minutes and adjustments should be made, if required. If, as in the melba toast example, the cause of the color unit drift is found to be oven temperature, then evaluation and correction should be made by a temperature sensor and controller, if available. In a manual

Figure 8-4.   Average of five successive samples taken at 12-second intervals.

process, the evaluation would be performed by an operator instructed to adjust the oven temperature if a reading is outside the process control limits. If the cause of color drift is found to be oven conveyor speed, then this would be the factor to be adjusted to maintain process control.

A number of sampling and adjusting schemes can be devised, depending on the process, the equipment, and the operators. For example, a 14-minute evaluation cycle could be established with the understanding that if the process drifts out of limits, as much as 7 minutes of production might be out of the color control limits.

Note that the process control limits determine the *obtainable* uniformity of product. In some instances, this degree of uniformity might be more than is required by the customer. In such a fortunate situation, the process need not be adjusted until the specification limit is reached, possibly reducing the cost of production.

The many uses for process capability studies include the following:

- *Process selection.* When several processes are under consideration, capability analysis will assist in selecting the most effective one.
- *Tolerance compliance.* Process capability analysis (PCA) will predict the extent to which the process is capable of holding.
- *Machine selection.* PCA can be used to assign machines to the classes of work for which they are most suited.
- *Selection and training of inspectors.* When a PCA has been prepared, the control limits provide reliable benchmarks to evaluate inspectors' ability to adequately measure quality.
- *Identification of troublesome steps.* In a series of sequential processing steps, PCA of each step may identity the one responsible for quality abnormalities. This reduces the research necessary to find a solution.
- *Research criteria.* Theories of defect causes may be readily evaluated by PCA during quality improvement programs.
- *Audit tool.* A series of PCAs performed on the same process will often disclose the frequency of process reviews required to ensure adequate maintenance, inspector instruction, operator performance, personnel supervision, analytical instrument calibration, and so on.
- *Setting specifications.* If a product is to succeed, its quality specifications must lie beyond the process capability limits.

CAPABILITY INDEX

One of the end products of a process capability study is the control chart showing the upper and lower control limits of a controlled process. The control limits cover a 6-sigma range. Another end product is the *capability index,* which compares the product specification range with the 6-sigma range. The index may be useful to those who prefer to express test results in numbers rather than words. It may also be valuable when a company has a large number of specifications and is constantly comparing process capabilities with control limits.

The calculations are simple:

$$\text{Process capability index} = C_p$$

$$\text{Capability index} = \frac{\text{Allowable process spread}}{\text{Actual process spread}}$$

$$= \frac{\text{Specification limits}}{\text{Process control limits}}$$

$$= \frac{\text{USL} - \text{LSL}}{\text{UCL} - \text{LCL}}$$

$$= \frac{\text{Tolerance}}{6\sigma}$$

These equations say the same thing: Subtract the lower specification limit from the upper specification limit, and divide the result by the difference between the upper and lower control limits (or six standard deviations). The resulting number determines whether or not the process is capable of meeting the specifications.

If the specification spread is the same as 6 sigma ($C_p = 1$), then approximately 0.3% of the production can be expected to be out of limits. If 6 sigma is less than 75% of the specification spread, then the process is generally considered acceptable and capable (i.e., $C_p = 4/3 = 1.33$.) In tabular form:

Capability index	Conclusion
1.33 or greater	Process is satisfactory.
1.00 to 1.33	Process is capable, but marginal as the index approaches 1.00.
1.00 or less	Process is unsatisfactory.

Example

Assume that the specification for a food product limits the ferric iron content to between 14 and 22 parts per million. A process capability study shows control limits of LCL = 15 and UCL = 21;

$$Cp = \frac{22 - 14}{21 - 15}$$

$$= \frac{8}{6} = 1.33$$

Therefore, this process is capable of meeting the specification. Figure 8-5 is a control chart illustrating this example.

To review: If a process is in control, and if all of the measurements of that process which lie between the lower control limit (three standard deviations below the average) and the upper control limit (three standard deviations above the average) also lie between the lower and upper specification limits, then the process may be considered *capable*. Figure 8-6 is an illustration of a capable process.

A method to avoid calculating the standard deviation of individuals data: Calculate the average range ($\bar{R}$) of differences of each successive pair of data points, and find the control limits for individuals using the following formulas:

$$UCL = \bar{X} + 2.66\,\bar{R}$$

$$LCL = \bar{X} - 2.66\,\bar{R}$$

Figure 8-5. Ferric iron content (in ppm).

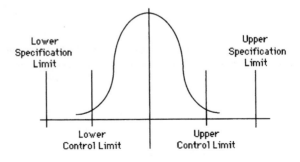

Figure 8-6. A capable process.

A shortcut to finding the standard deviation for individuals when a previously calculated control chart for averages is available: Multiply the UCL and LCL by $\sqrt{n}$, where n is the sample size used in the control chart for averages. The $\sqrt{n}$ converts average data to individual data. These individual control limits may now be used to calculate the capability index using the formulas shown on page 165.

The capability index discussion above refers to 6σ as the area between the upper and lower control limits. This should not be confused with the 6σ quality program popularized by an electronic manufacturing company. This organization was determined to reduce defects to parts per million, rather than the generally accepted 3 parts per 1000. In order to accomplish this, they started by placing the *specification* limits at 6 standard deviations on *either side* of the mean, shifting the *process* mean 1.5 standard deviations higher than center, and improving the process until the upper *process control* limit was within the specification limit. (Note carefully the italicized *process* and *control* verbiage in this explanation.) Perhaps Table 8-2 and the following discussion will make this clearer.

Research had determined that the process mean normally shifted higher by $+1.5\sigma$ for the company which evolved this quality control system. The units out of specification under these conditions would have been expected to rise to 6.8 per million; however it was theorized that because the lower control limit was also raised by 1.5σ, there would be zero defects at the lower level, and one-half of 6.8, or 3.4 defects, at the upper level. With this reasoning, the company identified the quality program as the Six-Sigma Quality Program, which produced 3.4 defects per million.

Because the program would be successful only if process improvements could be accomplished to fit within it, one might suggest other, perhaps simpler ways to accomplish the same defects per million goal. On the other hand, the numeric goal oriented system has motivated employees, and positive results have been obtained.

BENCHMARKING

The usual understanding of the benchmarking process is based on finding an external goal obtained from the best of industry practices. By investigating and in-

Table 8-2. **Relationship Between Capability Index and Defects**

Capability index	Specification limits	Units per million out of specification
1.0	$\pm\ 3\sigma$ or 6σ	2700
1.33	$\pm\ 4\sigma$ or 8σ	63
1.48	$\pm\ 4.5\sigma$	6.8
1.67	$\pm\ 5\sigma$ or 10σ	0.570
2.0	$\pm\ 6\sigma$ or 12σ	0.002

corporating these best practices into its operations, a company can expect to better serve its customers, and improve its processes and profitability. Definition of external benchmarking suggest continuously measuring and comparing a company's products and processes with those of industry leaders in similar products or services, and adopting the best of them with whatever modifications are necessary.

Direct competitors are unlikely to embrace this policy of sharing best practices, depending partly on the market share of each company. However, companies with different products or uncommon market niches may both benefit from frank discussions of each other's best practices. Other sources of information are trade journals, the Internet, business consultants, market research organizations, business and professional association meetings and publications, customers, and equipment vendors.

There are many differing benchmarking plans offered in the literature. Rather than recommending a single series of steps and rules for conducting a benchmarking program, it might be of greater value to consider some of the many questions which must be resolved before starting.

What management level should interview which host management level?
How is the benchmark target selected?
How can the host's information be tactfully verified?
Should the first project be a surefire success or a major goal?
Must contacts be performed with written questionnaires, or verbally?
Do host's measurements have the same meaning?
Are personnel qualified to perform, or are consultant's required?
Should informal contacts be made in addition to office visits?
May an undefined goal be used as a starting point?
Is a project schedule desirable—or possible?
Does the host's differing corporate culture account for success?
Are the units of measurement the same for the two companies?
Would information sharing violate any law?
How will use of new information be processed in the company?
Should benchmarking be conducted by a team or a department?

The techniques involved in benchmarking have been formalized by several companies and authors, and they all seem to have a central theme: Plan/Do/Check/Act (the Deming Cycle.) However, the terminology used in benchmarking steps varies: analysis, implementation, integration, measurement, identification, data collection, functional goals, establishment of performance levels, develop action plans, and monitor.

Not to be overlooked is a perhaps more valuable approach to benchmarking: internal process and product improvement made possible through examining the most successful principles and techniques which exist in one's own company. The discussion of benchmarking has been included in this chapter on process capability for that reason.

Internal benchmarking is an extremely effective tool and is usually welcomed by both production personnel and management, as it represents the best efforts and previous attainable successes found within the company. In essence, it is based on a simple principle: "We did it before and we can do it again." Perhaps most companies have repeatedly used a form of this technique without formalizing it with a title.

Although internal benchmarking can be successfully handled by a single engineer, production manager, researcher, or department head, it is more likely to be successfully implemented if a team is formed from various company operations. The increased success observed may be due in part to the possibility of contributions of more diverse ideas; but because the team includes members whose departments are affected by the final decisions, they would likely be motivated to contribute willingly and to actively implement the findings.

The principle of internal benchmarking is to select an operation which worked better than normal—one which exceeded the normally acceptable (and desirable) control limit for quality, cost, sales, personnel requirements, productivity, line changeover, absenteeism, consumer complaints, mistakes, scrap, production interruptions, accidents, billing, specifications (such as color, flavor, weight, dents, leakers, lumps, solubility, flow, granulation, and printing), or others. In a well-structured quality control system, this selection process can be as simple as forming a team to identify favorable outliers on process control charts and selecting the one which appears to have the greatest potential. In its simplest terms, benchmarking now proceeds to find out how the favorable outlier occurred, set up a test to duplicate those conditions, and, if successful, change the methods of operation to include the new conditions.

How does internal benchmarking differ from the normal principles of quality control and process improvement which have been used for years? Not by very much. The major change has been the introduction of a formalized structured system which includes (usually) team members from several departments. Each company can best decide how to structure the benchmarking process so that it will fit in with their culture. Following is a suggested framework for such a structure.

Form a search team to select a project area.
Establish a specific goal and major objective.
Select a team consisting of members affected by the goal.
Train team to follow specific procedures:
    Brainstorm goal to select specific project
    Establish resources necessary to achieve goal
    Determine data collection procedures
    Gather and analyze existing data
    Generate additional data as required
    Redefine goal and target
    Propose improvement plan (PLAN)
    Educate and train those involved with plan execution
    Implement plan (DO)

Determine effects of implementation (CHECK)
Apply corrections or modifications to implementation
Standardize procedures, specifications (ACT).
Prepare report on findings, action taken and benefits.
Publicize successes.
Review benefits and start a modified or new improvement project.

The major advantages of internal over external benchmarking are the availability of readily interpreted data, some of which might be proprietary, and the knowledge that the culture, environment, and operations are constant.

9   Process Control

Once the initial investigation of a process is complete (specifications and capability analysis), control charts can be constructed and evaluated to guide the continuous production of acceptable quality products. From the company's viewpoint, this is the quality control manager's most visible and useful function. When progressing from capability analysis to production quality control, use of single-sample measurements of variables are generally superseded by subgroups so that control charts for averages and ranges may be constructed. Procedures for establishing and maintaining these charts are described in Chapter 3. Our concern here is the interpretation of these charts.

In order to interpret a chart, it is essential to review a number of factors concerning its selection and construction. The function of charts used by a long-established company often erodes into a routine. If its original purpose is forgotten, a chart may have little impact on the control of quality. Under these conditions, the question, "Why do we sample at this point in the line?" might be answered with a shrug and the reply, "We have always done it this way." Many things change over time: New equipment added to the line might make some older measurements less significant, changes in raw products or packaging materials might not react to obsolete controls, and new methods on the line that critically affect quality might be overlooked by the existing quality control procedures. Periodic review of the principles on which the charts were constructed will avoid these problems.

Such a review should consider:

1. Training of the people currently involved with preparing the charts.
2. Assuring that the process location where the chart is generated is still a critical control point.
3. Determining whether or not this chart has an appreciable bearing on the customer's needs.
4. Judging the adequacy of the measurement system used from a standpoint of calibration for precision and accuracy, and state-of-the-art techniques.

5. Considering the possible need for an updated process capability study and control point revision, as dictated by either process improvement or specification change.

Control charts are a continuous "picture of the process" and, when properly constructed, provide the means of continuously producing product of uniform quality. When a single point falls outside the control limits, something is probably wrong. There are two key words in that sentence. *Something* might be the miscalculation or misplotting of the offending point, improper sampling due to stratification or ignorance, or data that have been edited. The other key word is *probably:* 0.3% of the observations of a controlled process can fall outside the control limits by chance alone. Because of these possibilities, a number of ground rules have been established by various quality control practitioners. Some say that a control chart need not cause alarm until two points are out of limits on the same side of the control limit. Others say three. Some systems require action when there seems to be a trend or pattern in the data. Other systems call for two standard deviations as a warning limit, or seven points all on the same side of the process average line. There are even some shortsighted (and probably short-lived) companies which use the ground rule: "No matter how many chart dots are out of limits, we are so far behind in production that we will not stop the line for anything short of a fire."

There are special cases where ground rules might be applied, but it is unlikely that a control chart will be misinterpreted *if every out-of-limits point were investigated, and if all other combinations of points within the control limits were ignored.* This statement applies equally to *X*-bar, *R, p, n,* and *c* charts. One of the greatest sins in attempting to control quality, and one of the major contributors to production costs, is attempting to control quality by adjusting or modifying when the process is already within the control limits.

With the understanding of the statement in the preceding paragraph, we examine below some special cases where peculiar chart patterns might be handled as exceptions to the rule. But first, let us discuss some general principles relating to data points that appear outside of the control limits.

Data on variables charts may first exceed the control limits in one of three ways:

1. If the average is out of limits but the range is not, there is probably a serious problem with the process. When this does occur, chances are that all of the observations in the subgroup are out of control (or nearly so).
2. If the range is out of limits but the average is not, the process may be on the verge of developing an oscillation. It can be expected to soon drive the average out of limits as well. It is likely that individual items are already close to or outside of the specification limit.
3. If both the average and the range are out of limits, something drastic has occurred, and it can be expected to become worse unless immediate action is taken.

When attribute charts (*p, np,* or *c*) exceed control limits, chances are that a substantial quantity of questionable quality products has been produced; if swift corrective action is not taken, then charts become useless scorecards, and reject/rework loads will build up.

When any chart shows a reading below the lower control limit, immediate attention is needed, but this does not always mean shutting down the line. In fact, exceeding the lower limit may be advantageous: fewer can leaks than permitted, lower bacteria count, less than the minimum moisture vapor transfer rate, unusually low bitter flavor characteristic, a total absence of contamination, or zero sediment. These and many other similar variables and attributes may all be very desirable in the product but are generally unattainable. When this type of outlier data occurs, the process may not necessarily need to be stopped; causes for the unexpected quality improvement may form the basis for an improved product and should be explored before they disappear. Once uncovered, permanent changes may be instituted and new process control limits adopted.

Of course, this might also be true for charts of some variables where exceeding the upper control limit could produce a more desirable quality product. Here, too, the causes should be explored with the goal of product improvement in mind.

CHART PATTERNS

1. Normal, In Control

The pattern in Figure 9-1 is random and within the control limits. No action is either needed or desirable.

2. Random Cycles

When random cycles (see Fig. 9-2) are found in the *X*-bar chart, the causes may be due to rotation of operators or machines, merging processes (such as feeding

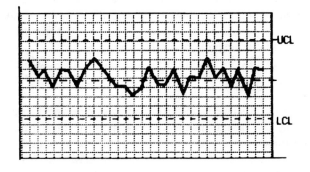

SAMPLE NUMBER

Figure 9-1. Normal, in control.

SAMPLE NUMBER

Figure 9-2.   Random cycles.

SAMPLE NUMBER

Figure 9-3.   Trend (down).

reworked material into the line, or alternating raw material and supplies), changes in environmental conditions (such as temperature or humidity) or worker fatigue. On the range chart, such random cycles may be due to inadequate maintenance of equipment, resulting in wear and intermittent operation. On p charts, cycles may occur from merging processes. In addition, failure to use the correct sampling plan for the chart in use can produce cycles.

3. Trend (Up or Down)

Trending in one direction (see Fig. 9-3) may occur in the chart for averages as one raw material is used up and another is introduced. Gradual deterioration of equipment, continuous change in the environment, or change in the production rate will also produce a trend. A trending range chart may be the result of gradual improvement or deterioration of a worker's skill, a production rate change, or a change in the number of components reaching the process. p Charts are similarly affected.

4. Abnormal Uniformity

All charts may show a reduced scattering of observations (see Fig. 9-4) when a superior well-skilled employee is present, when the process has been changed (intentionally or not), when a key step or key ingredient has been omitted, when defect-free raw materials are introduced, when unusually favorable environmental conditions exist, or when the wrong control limits are used. Nonrandom samples selected by an overzealous or improperly trained employee can result in uniform results. Samples tested on a malfunctioning instrument (such as a sticking scale, electronically insensitive meter, worn vibrator cam, or dusty colorimeter lens) or subjected to chemical or biological tests using outdated reagents may also show abnormally uniform quality. Range charts are particularly sensitive to conditions leading to abnormal uniformity.

5. Recurring Cycles

Charts that show cycles (see Fig. 9-5) may be related to activities that occur at repeated times (days, shifts, hours, or break periods) when changes in the process

Figure 9-4. Abnormal uniformity.

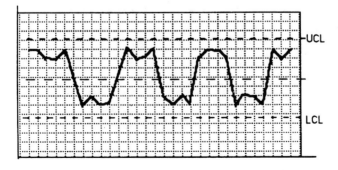

Figure 9-5. Recurring cycles.

or the environment occur regularly. The changes also may be due to workers, supervisors, inspectors, scheduled maintenance, ambient temperature or humidity, lighting, or production equipment rotation. Nonuniform cycles may be traced to occasional equipment stoppages for adjustment or repair, which may produce new quality levels when restarting. Cycles occur from time to time on range charts due to these causes, but X-bar and p charts are more likely to uncover cyclical quality problems.

6. Excessive Variability

When found on the chart for averages, excessive variability (see Fig. 9-6) is most frequently caused by overadjustment of a process. When an operator adjusts the pH downward simply because it has been above the average for a few readings, chances are the pH will shortly drift below the lower control limit. Blending production from two or more lines may drive the range chart out of limits (e.g., drawing samples of pouches for length measurements produced on blended lines, where one line is at the minimum length and the other is at the maximum). Malfunctioning equipment and intermittent use of variable raw material (such as feeding into the line raw material that is of variable quality or nearly out of specification) may produce peaks and valleys on the range chart and the p chart. A malfunctioning inspection device may also produce apparent wide variations.

7. Wild Outliers

When an occasional control chart plot lies far outside the limits with no apparent cause (see Fig. 9-7), it is generally the result of a quality testing error. Such errors include mathematical calculation, plotting, sample size, sampling procedure, testing, sample contamination, and so on. Problems associated with the production line are rare and generally nonrepetitive—an interruption of power or gas supply, a sticking solenoid or chattering relay, or, possibly, an omitted operation. Samples

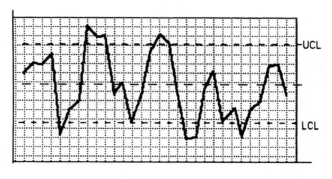

SAMPLE NUMBER

Figure 9-6.   Excessive variability.

SAMPLE NUMBER

Figure 9-7.   Wild outliers.

SAMPLE NUMBER

Figure 9-8.   High/low.

removed from the line and inadvertently returned to the wrong part of the line can also produce an otherwise unexplained outlier.

8. High/Low

X-bar and R charts will often show extreme values with few in between (see Fig. 9-8) when two sources of raw materials are used, one at the upper end of the quality specification and the other at the lower end. As the variable supply reaches the production line, the process swings rapidly from one end to the other, rarely pausing in the center. Since product defects would tend to average out, p charts might not be affected by two source materials. However, nonrandom sample selection could show high/low p-chart values. The blending of product from two dissimilar lines or two operators with varying techniques may be responsible for this unusual chart. The output of twin-pack packaging machines may produce units which are within—but at opposite ends of—the control limits. Depending on which side of the machine the sample is taken from, variables such as gas content, fill weight, package dimensions, dust content, seal strength, and registration may appear at either end of the control chart—but rarely in the middle.

These explanations for control chart patterns are necessarily incomplete. Considerable detective work is required to explain most out-of-limits data. Most answers are found on the production floor (or at the raw materials supplier plant); as the relationship between quality control personnel and operating personnel strengthens, answers and solutions to production quality problems are found more readily. Sometimes causes are strange and require a great deal of ingenuity to uncover—vibrations from the freight elevator upsetting the line check-weighers, burnt paper ashes strewn over the oven contents during the Kosher blessing, static buildup during thunderstorms causing blocking of plastic films on sealers, or conscientious line inspectors who want the company's record to look good and eliminate out-of-limits data.

USING THE CONTROL CHART AS A QUALITY MANAGEMENT TOOL

What does one do with the charts at the end of the shift—or day, or week? Something can be said for posting the charts on the production floor as they are prepared so operators can see how well they are doing. Continued operator interest is unlikely though, unless some kind of attention-getting technique is included. One could post the name of an outstanding operator for the week, prepare a scorecard comparing shifts (can be risky), compare the current month with the same month last year, or display a thermometer of accomplishment. All of these ideas lose effectiveness with overexposure, however, and some are difficult to discontinue after they have served their purpose, because a few employees will remain attached to them. Any such plan should be carefully considered before obtaining approval of management; competitions can cause jealousy, animosity, or even dishonesty if improperly administered. However proper use of a well-planned motivational quality tool is recommended if it is used infrequently and has a worthwhile message.

At one time or another, copies of control charts should be sent to everyone in the company who is even remotely concerned with product quality. Subsequent charts should not follow unless they are requested. If no requests are forthcoming, the charts must be made more colorful, informative, and interesting. Then they should be sent out again. Notes and comments should be included, showing solved problems, quality improvements, productivity gains, and cost reductions. Most important, credit should be shared (deserved or not) with production, engineering, accounting, purchasing, and marketing departments whenever even remotely possible. Suggestions should be solicited.

For those few managers who ask for daily charts, their interest should be rewarded with brief summaries, weekly and monthly reports, and short notes about other industries' quality problems and solutions. Charts should be informative, and they should be spiced and embellished with cartoons or photographs of quality improvements so that the recipients actually look forward to the next issue. As an additional courtesy, provide managers with informal tours of the production floor where they can watch a chart being constructed.

We have left the most important point for last. How does one actually use these charts to control quality effectively? That is, what physical steps must quality control personnel take to make the charts work? When an observation exceeds the control limits, the line should be immediately stopped (if permitted by management) and all production since the previous observation set aside and clearly labeled with a HOLD or NOT TO BE SHIPPED or REWORK tag. Once the causes for the out-of-limits condition has been identified and corrected, the line may be permitted to start up again, and follow-up paperwork providing instructions for isolating and disposing of the defective material should be completed.

The alert reader might have noticed the parenthetical phrase, *if permitted by management,* in the preceding paragraph. For many, this may be difficult to understand, but there are companies which permit only production supervisors to stop production lines. It is still possible to operate a quality control system effectively under these conditions, however, for it is unlikely that any management would also forbid the quality control department to isolate and hold any defect, whether it be raw material or finished product. The out-of-limits material can be isolated at the end of the production line and then held for corrective action. Eventually, the production department will realize that all production is headed for the rework or scrap area—and will continue to be until the line is stopped and the offending condition corrected. Needless to say, one or two such experiences will lead to a revision in management philosophy, and quality control personnel will eventually be authorized to stop defective production.

Eliminating out-of-limits materials as they are discovered in the process, through the active use of control charts and immediate corrective action, is at the heart of quality control; it is the means to reducing defect production and scrap, increasing productivity, and—most important—lowering costs. Some quality leaders have hailed this procedure as a newly discovered principle and provided it with such labels as statistical process control, zero defects production, process quality assurance, or other. So be it! Regardless of title, this procedure works well, and has for many years. Properly administered, it can be used successfully in any food manufacturing process.

10   Sensory Testing

Many think of sensory testing primarily as a research tool. It is widely used in product development to match products with competition, to determine consumer acceptance, to improve products, and to conduct shelf-life studies (storage stability). But sensory testing and evaluations are used more routinely for quality control functions such as raw material control, evaluation of new sources for raw material, process control (at every step of the process), evaluation of process change, product grading, and multiplant quality coordination.

Ask any healthy person if he or she can compare food flavors, and chances are the response will be yes. Ask, "Why are you qualified as a flavor judge?" and the answer will be something like, "Because I have tasted foods every day of my life." The only reasonable conclusion to be drawn from these answers is that this individual is capable of judging food flavors as pleasing or displeasing to himself—but not necessarily to anyone else.

Sensory evaluations of food are concerned with much more than flavor. Texture, color, size, density, aroma, and many other factors have sensory values which are often vital to the quality acceptability of foods. The threshold level at which food characteristics can be detected, or the levels at which they can be differentiated, need to be understood. Then there are preference values dictated by cultures, ethnic practices, age, geographical area habits, and so on. When we refer to sensory testing, we are concerned not merely with one person's evaluation of flavor, but rather with a vast field of sensory perceptions which require sorting out. Fortunately, a great deal of research has been devoted to this field, and sensory testing programs can be established at an almost infinite number of degrees of complexity.

The scientifically trained person finds it difficult to understand why a clear preference for one food item over another does not lead to instant success in the marketplace. The significance of preference values is clearly demonstrated by a test conducted many years ago by the U.S. Army Subsistence Research and Development Laboratory. A consumer panel of nearly 300 subjects almost unanimously rated a "new" flavor of ice cream as "excellent." When the same golden-colored ice cream was presented to the same group 3 weeks later, but now labeled as "carrot-flavored," the ratings were "unacceptable" or worse. The lowest rating was "revolting." The art of marketing may start with the logical findings of a sensory panel, but presenting a winning product to the marketplace requires techniques beyond the scope of quality control and will not be discussed here.

In 1975, the Institute of Food Technologists agreed on a definition of sensory evaluation: "a scientific discipline used to evoke, measure, analyze, and interpret reactions to those characteristics of foods and materials as they are perceived by the senses of sight, smell, taste, touch, and hearing" (1981).

THE SENSES

Taste

By far, this is the most commonly evaluated characteristic of foods. Taste can be classified into four basic flavors: sweet, salty, sour, and bitter. (The thousands of characteristics which we loosely call *flavors,* such as onion, strawberry, mint, and butter are better described as *odors.* If one is blindfolded and fed a small piece of onion while squeezing the nose shut, it will most likely be identified as apple. The characteristic onion *flavor* is actually onion *odor.*)

Odor

There are thousands of odors associated with foods. To date, a simple classification of odors acceptable to the food industry has not been devised. In a paper delivered to the 1982 Institute of Food Technologists Annual Meeting by Boelens and Haring of Naarden International (and others), the following 30 odor classifications were utilized to compare odors of a single group of products:

Fresh	Green	Sourish	Tart	Citrusy
Watery	Metallic	Floral	Fatty	Aldehyde
Vegetable	Lavender	Coniferous	Minty	Medicinal
Fruity	Honey	Buttery	Animal	Erogenic
Sweet	Aromatic	Anisic	Spicy	Powdery
Dusty	Earthy	Smoky	Woody	Balsamic

Each of these classifications was described by standardized chemicals, natural flavorings, oils, or other substances, and may not be universally accepted as descriptive. For example, *minty* was defined as "like peppermint oil," *buttery* as "diacetyl," and *aromatic* as "vanillin." For purposes of the study, however, these descriptors proved satisfactory.

Combinations and Interactions of Taste and Odor

Here we find an infinite number of possibilities. No attempt has yet been made to categorize them for all applications, but simple definitions can be used successfully.

Heat, Cold, Pain, Touch, Sight, Sound

At first, these terms might sound farfetched, but it is difficult to find a description more easily understood for the sensation of hot pepper than *heat*. Likewise, some wintergreen mixtures which are *colder* than others. *Touch* might be interpreted as texture, greasiness, or sound. Yes, sound can play a significant part in the perceived quality of some food products—the crackle of cold cereal in milk, the snap of chewing nuts, the hiss of carbonated drinks. Sight is so critical to flavor perceptions that some tests may be conducted blindfolded, and many tasting booths are equipped with special lighting to accentuate or deaden color effects. When meat is presented as an appetizing steak surrounded by its own juices and with barbecue grating marks on the surface, flavor perception is totally different from the same meat presented as a ground, dull-brown, shapeless mass.

The physiological aspects and nomenclature of sensory perception are obviously quite complicated. The examples included here show how a "yes" answer to the question of whether one is qualified as a flavor judge is not a criterion for selecting a sensory panelist. For a more detailed examination of stimuli and the physiological background to sensory evaluation, see *Basic Principles of Sensory Evaluation,* American Society for Testing and Materials Special Publication No. 433 (ASTM, 1968).

SENSORY TESTING METHODS

Selection of a sensory test method is dependent first on the type of information required, and second on the precision required of the results.

Common questions to be answered are as follows:

How do you like this product?
How do you like the color of this product?
Are these two products the same?
Are all of these samples identical?
Arrange these products in a column with the strongest at the top and the weakest at the bottom, with intermediate ones in order of their strength.
Arrange these products in a column with the one you like the most at the top, decreasing until the one you dislike the most is at the bottom. Identify the one you neither like nor dislike.
Which samples have vanilla flavor?
Which two of these three samples are alike? Are they better flavored than the single sample?

Once the questions are determined, a choice of test will usually be found among the following methods.

Monadic.   The tester evaluates a single sample, based on past experience only. This method is useful when an experimental product contains a major departure from similar products on the market—such as lime-flavored apple juice or hamburger containing tiny julienne potatoes.

Paired comparison.   A tester compares Sample A with a somewhat similar Sample B, and then describes specific perceived differences. Results are analyzed using the two-tailed test, since either sample could be preferred. In contrast, the assertion that one mean is greater or lesser than another (more salt, less sugar, stronger flavor, etc.) requires a one-tailed test.

Threshold.   A tester identifies which of several samples (usually five or more) do not contain a given characteristic and which do. The samples are prepared by adding some characteristic to a series of "blanks" in small incremental quantities.

Dilution.   This test determines the lowest concentration at which characteristics can be detected. It differs from the threshold test by use of complete product at different strengths; threshold tests are generally concerned with a single component in a bland carrier.

Triangle.   Two samples are identical, and one is different. The tester identifies the pair. Is the pair preferred for some designated characteristic?

Double triangle.   In this test, there are two sets of triangles, not necessarily with the same pair in each. Testers identify the pairs and their preferences.

Duo-trio.   Two samples are identical, and one is different. One of the identical samples is presented as a control, and the tester is asked to identify the sample which differs from the control.

Rating scale.   Testers apply a rating to each of a coded group of samples. The scale may be a line with *good* at one end and *poor* at the other; or adjectives such as *fair, OK, good,* and *excellent;* or numerical scales representing successive levels of quality; or other scalar devices. This is also referred to as an *optical intensity scale.* Results may be analyzed using analysis of variance.

Magnitude estimation.   This is similar to the rating scale, except that the tester is asked to apply quantifying scales to some characteristic of a series of samples. For example, the strength of a component might be assigned values from 0 to 100 by the panelist, the values being of any perceived interval, not necessarily uniform.

Ranking.   The tester is asked to place samples in order from *most* to *least,* or *best* to *worst.* This test differs from the rating scale in that no suitable scale may be available. It works best for preliminary screening and when differences are fairly obvious, requiring little time for evaluation.

Attribute analysis.   A specifically trained tester is presented with a single sample and is asked to list all attributes defining the quality of the sample. For example, the list for a candy bar might include chocolate flavor, sweetness, size of nuts, darkness of color, crunchiness, creaminess, grit, chocolate aroma, sour aroma, bitterness, and aftertaste. Results from 6–12 panelists may be analyzed for significance by using analysis of variance.

Hedonic scale. This is similar to the rating scale, except the criterion here is the extent to which the samples are liked. No attempt is made to direct responses or define the meaning of scale categories. The scale is presented in nine categories: *like extremely, like very much, like moderately, like slightly, neither like nor dislike, dislike slightly, dislike moderately, dislike very much,* and *dislike extremely.* When less-experienced panelists are used, the *moderate* categories are eliminated. When a panel consists of children, the following descriptors are suggested: *love it, like it a lot, like it a little, it's just so-so, dislike a lot,* and *hate it.* Hedonic scale testing is most useful when conducted with 40–50 nontrained office personnel. If conducted outside the company, more than 100 consumers should make up the panel. Results may be analyzed using analysis of variance.

Flavor profile. Five characteristics are identified by the subject: character, intensity, order of appearance, aftertaste, and fullness. *Order of appearance* refers to the time sequence at which the various aroma or flavor components appear. *Fullness* refers to the overall impressions of blending of quality components and the appropriateness of the factors to the product. This method is not subject to statistical analysis. After panelists independently evaluate one to three samples, they then report to a panel leader in open discussion. (This test method can be adapted for texture or other characteristics.)

Texture profile. Three classes of characteristics are identified and described: initial perception on first bite (hardness, viscosity, fracturability), masticatory perception during chewing (gumminess, chewiness, adhesiveness), and residual perception during mastication (rate of breakdown, type of breakdown, moisture absorption, and mouth coating). Statistical analysis is not appropriate to this method. Individual judgments are discussed to arrive at an average rating for each of the three classes.

Routine testing. When a large number of production samples is presented to a quality control technician for comparison to the quality standard, the usual procedure is to taste them in rapid succession, referring to the standard only when there is some doubt about the characteristics of an individual sample. Generally, the products are smelled, tasted, and chewed (if applicable), but not swallowed. If there are lingering characteristics, a rinse is recommended between samples. In the event of a suspect sensory characteristic, a second opinion is recommended. When the number of flavor variables is relatively small (such as with lemonade, coffee, or flavored pudding), as many as 30 samples can be evaluated at one time. When the number of flavor or texture variables is large (such as with beef stew, mixed vegetables, or fruit pie), it is unlikely that even an experienced technician can handle more than six samples at a time.

This formidable list of available tests raises many questions. There are no universal answers, but the following guidelines might be considered.

Fatigue and Frequency of Testing

In-house panels consist of office, plant, and laboratory personnel who have responsibilities, interests, and time constraints which may pressure them to rush through the sensory tests. Obviously, this will affect their reliability. On the other hand, some subjects enjoy the break from the monotony of their routine. With the exception of those whose duties require continuous sensory testing, such as quality control laboratory inspectors and technicians, it is wise to restrict the number of tests in which a subject participates to no more than one per day, and preferably only one every other week. At a single sitting, the product will determine the limitations as much as the testers' abilities. Weak- or mild-flavored products can be presented in as many as three simple tests without fatigue, but strong or harsh flavors can dull the senses very quickly. A single test of, say, spiced olives will end the day's testing for any panelist. Alcoholic beverages present very obvious limitations to the amount of product which can be consumed before test results become suspect.

Performance Grading

Many panelists will hand in their score sheets with the question, "Did I get them right?" For some tests, there are no right or wrong answers, but when a participant selects a pair from a triangle test or grades five aged products for staleness, there is no harm in disclosing the accuracy of responses. In fact, there are some advantages. There is a certain amount of pride in knowing that one has the ability to discern the correct answers, and this might motivate testers to continue to concentrate on future tests and look forward to participating again. Knowing that there are correct responses may help some people avoid guessing at the answers. For those who missed the correct answers, perhaps they will try harder on the next test.

Blind Versus Open Testing

Exploratory tests may be performed open; that is, the panelists may be told which sample is the standard, what the other samples are, and the purpose of the test. It instills a measure of confidence and speeds up the testing. In blind testing, where unbiased data are essential, some information can—and perhaps should—be made available to the tasters. For example, presented with five oranges and asked to evaluate them for appearance (without further explanation), some panelists might look for size differences. The test might have been designed to evaluate the effectiveness of various levels of controlled atmosphere storage, or different temperature storage, or irradiation, but the testers might totally miss rating the samples on the basis of color, absence of mold, shriveled skin, or odor—any of which could have been the major goal of the test.

Order of Presentation

When samples are presented one at a time for evaluation, the order of presentation becomes significant. Each sample is subconsciously related to decisions about the preceding sample. This *contrast effect* varies among subjects, and neither experience nor training will eliminate it. It can be reduced by changing the order of presentation to each subject and, if possible, changing the order in repeat testing as well. Even when samples are presented simultaneously, as in triangle tests, *position error* can affect results. For that matter, the first sample selected in a triangle test will bias evaluation of the next two.

Temperature Control

For difference testing, most products should be presented at room temperature. For preference testing, they should be at the temperature normally served. If unsure of the effect of temperature, tests may be repeated at various temperatures and the results compared.

Reference Standards

If the company's product A is to be rated against similar competitive products X, Y, and Z, should A be identified as the company's product for comparison purposes? Generally, the answer is no. Most employees on in-house testing panels have a certain loyalty to the company and tend to rate their own product as superior whenever it is identified as such. After the test is completed, however, there probably is no harm in informing the tester. When the development of off flavors in storage is a concern, presenting testers with an identified fresh standard is desirable. This allows the tester to concentrate on the possible degradation of product quality, rather than expending unnecessary time and effort trying to establish which sample is the standard.

The techniques for preparing a true standard for a taste test are often difficult to establish. When comparing the browning reaction in sliced apples, the standard selection is fairly simple: nearly any fresh apple slice from the same variety can be used. But in storage tests where staleness development is the key characteristic to be tested, it may be extremely difficult to present an identical (but fresh) sample as a standard. Potato chip samples selected off the production line cannot be used as a standard for test samples that have been stored under various gas mixtures for various lengths of time at various temperatures, because current production may represent a different type of potato with different flavor, texture, and color characteristics. The same is true in testing coffee for staleness development during storage: The current production might consist of a different blend with significant flavor differences. In general, the best standard would be one selected at the same time the storage test sample was prepared, hermetically sealed, and held frozen at $-15°F$ until required for the test. Obviously, this will not work for all products because of physical changes which take place at freezing temperatures. Where no satisfactory standard can be provided, it is best to offer none rather than provide a sample with instructions to ignore the mushy texture, excess fluid, or unusual appearance. Such instructions would lead to questionable evaluations.

When it can be demonstrated that the reference standard changes with time, it may be possible to provide a continuous series of updated reference samples. For example, let us assume that a reference sample, stored frozen in a hermetically sealed container, changes imperceptively for the first four months but has been proven to show a gradual change after that period. At the end of three months, an updated sample can be prepared by selecting a series of samples directly from the production line and comparing them to the thawed standard by use of triangle tests. When a new sample has been found to be indistinguishable from the thawed standard, it may then replace the present standard (which should be discarded).

In addition to the stored product standard and the fresh product standard, there are three other standards in general use:

- *Written standard.* A documented flavor profile may suffice as a flavor standard for relatively uncomplicated generic types of products. It may be made somewhat more definitive by including charts of chromatographic or spectrophotometric peaks. Although this may function satisfactorily for awhile, its effectiveness becomes questionable as sensory judges move on and are replaced by others who are less familiar with the sensory objectives of the original product defined by such a document.
- *Photographic standard.* These standards are quite effective for defining physical characteristics (colors, blemishes, defects, size, proportions, and so on), provided the photographs are suitably protected from changes.
- *Verbal or mental.* Some flavors and textures can only be communicated to others through experience. Examples of these include wood, tallow, hide, wine, cardboard, nutlike, straw, acidic, and gummy. Training most healthy people to recognize these characteristics is not particularly difficult. The main problem is defining the extent or strength of these characteristics in a given sample. A second difficulty is training panel members to uniformly rate the degree or intensity of these characteristics. By using larger panels for these inexact measurements, the average reaction becomes more meaningful.

TYPES OF PANELS

Experts in the sensory evaluation field tend to treat the subject as if it were directed to large companies with vast financial resources. This leaves the quality control supervisor for a small company in a dilemma; his sensory panel consists of himself, Mary Smith in purchasing, and the plant manager. Occasionally, the company owner will sit in, with comments from his wife regarding the desirable flavor characteristics for the product in question. How can the quality control supervisor provide separate panels for the many sensory applications that arise? He cannot, but armed with information regarding the functions of specialized panels, he should certainly be able to provide at least the basics of sensory quality control to his company. In many instances, this might be all that is needed. Where specialized assistance is really required, it is available in outside consulting organizations or even in noncompetitive food processing companies.

Screening Panels

These are generally found in the research department, where an informal decision is reached to continue development of an idea or to try again with a different direction. Such panels should be staffed with members competent in the area under consideration; as fine-tuning is not required at this point, however, reasonable experience with the product class is generally satisfactory. Other screening functions may be carried out in the quality control laboratory for a variety of needs. In sorting rejected material, there is usually only one major factor to be evaluated (the one which caused the rejection), and it is customary to use semiskilled personnel to screen out the offending material. Another quality control use of the screening function is to discover unwanted sensory characteristics in the process stream at each critical step. This is accomplished by use of a highly trained individual (preferably with a backup trainee) who samples the process every 15 minutes (or other time interval, or batch) to identify any undesirable sensory characteristic at the earliest step possible and at all subsequent steps. This would include, but is not limited to, unusual color, texture, flavor, mold, foreign materials, and so on.

Expert Panels

These are groups of specially trained individuals whose function is to make fine sensory distinctions between products or evaluations of single products. In addition to being used by the research department, experts are often routinely used by raw material purchasing departments to evaluate both offerings and receipts. Frequently, they will check the finished product to determine if their selection of raw material withstood processing without showing undesirable characteristics. It is relatively unusual for a quality control panel to maintain a special expert panel. In the rare event that the routine quality control sensory panel finds an unidentifiable characteristic, the product can be sent to the expert panel elsewhere in the company for further analysis. If none exists, it can be sent out to an independent laboratory (or, in some cases, to the material supplier).

In-house Panels

The term *in-house* merely identifies the panel as consisting of company employees who may or not be specially trained, and who may or may not regularly participate in sensory testing. Frequently, an in-house panel is one which evaluates specific products (new developments, critical complaints, competitive comparisons) as an inexpensive way to determine if outside technical assistance is required. An in-house panel may occasionally be used as a miniature consumer test, providing rapid and tentative results when broad-scale consumer tests are being conducted simultaneously by outside marketing organizations.

Informal Panels

In addition to the research and control panels which use relatively rigorous procedures in conducting their functions, companies may also have a number of informal groups who meet periodically to "see how things are going." An example might be the monthly sales or marketing meeting, where samples of the company's major products can be compared to those of the competition. The procedures are usually casual, with prepared samples being offered to each participant for comment. Results may or may not be formally reported to other departments. Under these conditions, major product differences discovered are liable to be viewed as catastrophes, and either the research or quality control department quickly becomes involved. The same type of informal competitive sampling may be performed by any concerned department—research, quality control, purchasing, and others. Because these tests are similar to the actual treatment a product might receive by consumers, some surprising characteristics are occasionally revealed. For example, a soup-mix pouch may be difficult to open because the tear strip is not clearly identified. The countersink on a vegetable can may be so excessive that a hand can opener will not work at all. Reclosing a partially used frozen food package might be difficult. Reheating a gravy product might cause curdling. A tea bag string might tear loose in hot water.

Acceptance Panel

This is a rather loosely defined term that describes the end result of any number of sensory panels—acceptance or rejection of a sample. In quality control, an acceptance panel is one that can be used in audit functions, daily routine sampling, or sorting rejected material. In a manner of speaking, however, screening panels and in-house panels are also acceptance panels.

Interplant Panels

Some companies with more than one plant might wish to dignify the function of those quality control panelists who are responsible for maintaining identical product quality characteristics in all plants with a special title, interplant panelists. At best, it might be a part-time duty of those responsible for routine sensory testing. This is not to belittle the function—it is of prime importance and requires special quality control efforts. The greatest contributing factors to multiplant or even international quality control is the development and maintenance of strictly enforced standard methodology, including a communication network. Interplant panels are an integral part of the methodology.

Outside Panels

Outside assistance exists in many forms. Marketing and advertising companies frequently maintain panel facilities for advertising copy, label design, flavor testing, and focus groups. (Focus groups are panels selected or maintained by market research organizations to explore existing or new concepts concerning products. The groups are carefully selected to represent the potential consumer of the product, and the evaluations conducted by the panels are carefully monitored and directed by a professional group leader.) In addition to marketing and advertising organizations, outside assistance is available from consultants and commercial testing laboratories. Although the client company has the final word on how the studies are to be conducted and analyzed, it is wise to allow these outside facilities to apply the expertise they have developed over the years.

SELECTION AND TRAINING OF PANELS

In order to obtain meaningful evaluations from a panel, members should be selected and trained with care. Uninformed prospective panelists from within the company consider sensory panel assignment as a fun-and-games diversion from the routine of their regular jobs. Although this attitude must be modified, there is little to be gained by treating the assignment as a chore. Participating in panels can be a pleasant experience, but it must not be treated casually. Occasional light humor in the panel room is acceptable, but by no means should it be encouraged.

As a guide, an in-house panel of 8–10 members can be selected from a group of 15–20 volunteer employees. After two or three training sessions, the six most reliable and consistent candidates should be identified; others can be held in reserve as needed. The training sessions should include techniques in which the panelists follow well-defined procedures:

1. Rinse mouth with room-temperature water.
2. Sip (or nibble) a sample.
3. Chew (if required) and roll over front, sides, rear of tongue.
4. Spit sample into cup provided (do not swallow).
5. Record observations as instructed.
6. Rinse mouth until flavor sensation has disappeared.
7. Repeat for next sample.

The tests used for training should start with a ranking test in which a company's product is presented in four levels of flavor strength, sweetness, saltiness, sourness, firmness, or some other sensory characteristic appropriate to the product. Differences between the samples should be substantial for the early training sessions. Panel members should then receive training in the other types of test methods, starting with triangle testing. It is best to conduct the training program on consecutive days for at least a week. During these training periods, terminology can be gradually introduced as examples appear. Frequently, panelists will suggest descriptive terminology other than that commonly used, and this should be encouraged—and the new terms adopted if they clarify perception. Silence should be stressed during the training period; subconscious grunts, groans, or snickers may affect the judgment of other panelists.

The ability to duplicate results is as important as sensitivity to sensory stimuli, and a series of replicate tests must be conducted. Record keeping becomes voluminous at this stage, and the use of a computer and some sort of grading system should be devised.

Finally, some system of retraining should be considered. If tests are run frequently, panelists might drift away from standard procedures. An occasional refresher course in the form of a rigged test where participants openly discuss their findings might reveal developing weaknesses in the system which need to be corrected.

Figure 10-1 is a model for use in sensory panel testing for preference.

Figure 10-1.   Preference rating form.

11   Net Content Control

Controlling net content of food packages has two goals: First, production must fall within government limits for net package content; and second, overfill must be optimized (which really means *minimized*).

Federal regulations for net content of foods are covered by the U.S. Department of Health and Human Services, the FDA, and by the USDA Food Safety and Inspection Service. Alcoholic beverages are controlled by the U.S. Department of the Treasury, Bureau of Alcohol, Tobacco and Firearms. Nonfood consumer commodity net content regulations are covered by the Federal Trade Commission, and pesticides are under the jurisdiction of the Environmental Protection Agency.

Regulations dealing with net content control for food are less specific than one would desire. For example, the Code of Federal Regulations 21, Food and Drugs, Section 101.105 requires that

> The declaration of net quantity of contents shall express an accurate statement of the quantity of contents of the package. Reasonable variations caused by loss or gain of moisture during the course of good distribution practice or by unavoidable deviations in good manufacturing practice will be recognized. Variations from stated quantity of contents shall not be unreasonably large. (FDA, 1989).

The USDA (CFR 9) Meat and Poultry Act, Section 317.2, requires that the label "shall express an accurate statement of the quantity of the contents of the container exclusive of wrappers and packing substances" (USDA, 1988).

Neither of these regulations defines the word *accurate*. Does it mean *close to, at least equal to,* or *average?* Nor, for that matter, is the meaning of *unreasonably large variations* clarified. As a general practice, the American food industry overpacks, but according to the letter of the law, an extra ¼ ounce of product over the label claim does not comply with the requirement that the label "express an accurate statement of the contents." To further complicate the issue, each state has the power to enforce compliance with its particular weights and measures laws, usually through county bureaus of weights and measures.

Individual company policies vary in their attempts to comply. Some packers with "Zero-Defects" Programs may state that no product shall be underweight (although overweights might also be considered defects). Some managers might have 5% underweight and 95% overweight as the goal, or 10/90, or 20/80, or even 50/50. Others may have a more scientific approach. There are probably companies who have no policy at all regarding underweight–overweight ratios.

Fortunately, there is a way out of this confusion. The U.S. Department of Commerce, National Institute of Standards and Technology (formerly the National Bureau of Standards) has issued *Handbook 133* titled "Checking the Net Contents of Packaged Goods" (NIST, 1988). It replaces *Handbook 67,* "Checking Prepackaged Commodities," which served as the basis for most state weights and measures regulations. Gradually, states have adopted *Handbook 133* as the official procedure for evaluating net contents. Most states have also adopted the portion of the National Conference on Weights and Measures Model State Packaging and Labeling Requirements (from *Handbook 130,* 1995), which is related to averages and variations. It requires that the average quantity of contents in the package of a particular lot shall at least equal the declared quantity, and no unreasonable shortage in any package shall be permitted. It also requires that variations from the declared quantity shall not be unreasonably large. Maximum allowable weight variations are defined in *Handbook 133.* The National Institute of Standards and Technology has also published *Handbook 44* (1995) titled "Specifications Tolerances and Other Technical Requirements for Weighing and Measuring Devices." All 52 states have adopted *Handbook 44.*

Handbook 133 separates sampling plans into two categories. The plans in Category A (see Table 11-1) are used by weights and measures officials when the

Table 11-1.   Sampling Plans for Category A and Category B

Lot size no. of packages in lot (N)	Sample size no. of pkgs. in sample (n)	Tare sample size no. pkgs. to be tared (n_t)	Allowable no. of package errors exceeding the MAV*
CATEGORY A			
30 or less	all	2	0
31–800	30	2	1
801–2,000	50	5	2
2,001–5,000	80	5	3
5,001–15,000	125	5	5
15,001 and greater	200	10	7
CATEGORY B			
Up to and including 250	10	2	0
251 and greater	30	2	0

*Maximum allowable variation for individual packages; applies only to underfills (shortages).

severity of the consequences for the packager or retailer of a lot not passing the test is relatively great. If an official anticipates a possible regulatory action (court appearance), then Category A is used.

Category B plans are used as audits for screening purposes. In the event an audit reveals shortages, a federal or state agency might elect to follow up immediately with a Category A sampling. The plans resemble Military Standard 105E sampling plans for sizes over 50, but the decision criteria are different.

Maximum allowable variations (MAV) are listed in *Handbook 133* for packages labeled by weight, volume, count, and area. The number of sample measurements exceeding the MAV may not exceed those listed in the sampling plans for the appropriate category. The tables are fairly detailed. A few entries are shown in Table 11-2 as examples of MAV for individual packages labeled by weight.

For Category B, the sample average must be equal to or greater than the label weight. For Category A, the sample average must be equal to or greater than the label weight minus the adjustment for sampling error. Sampling error for *Handbook 133* is defined as two standard deviations from the mean. An additional allowance for moisture is calculated if a regulatory agency has assigned a value, and this allowance is subtracted from the "nominal gross weight." Later editions of *Handbook 133* have slightly revised this statement to read: "Allowance must be made for moisture loss during the course of good distribution when State or Federal regulations provide for them."

Lot sizes are defined for three locations:

- *Retail store:* A lot consists of packages with identical labels.
- *Warehouse:* A lot consists of packages with identical labels and with the same manufacturer's lot symbol or code.
- *On-line:* A lot consists of packages with identical labels and manufacturer's code and should not exceed one uninterrupted production run in excess of 1 hour.

Table 11-2.   Maximum Allowable Variations

Avoirdupois units			Metric Units	
	MAV			
Labeled weight (pounds or ounces)	Decimal pounds	Fractional ounces	Labeled weight (grams)	MAV (grams)
0.12+ to 0.18 lb. 1.92+ to 2.88 oz.	0.012	$\frac{3}{16}$	54+ to 82	5
0.58+ to 0.70 lb. 9.28+ to 11.20 oz.	0.032	$\frac{1}{2}$	263+ to 318	15
0.94+ to 1.08 lb. 15.04+ to 17.28 oz.	0.044	$\frac{11}{16}$	426+ to 490	20
1.40+ to 1.54 lb.	0.056	$\frac{7}{8}$	635+ to 698	25
1.88+ to 2.14 lb.	0.070	$1\frac{1}{8}$	852+ to 971	32

Samples are removed from the lot by use of random numbers selected from over 30 pages of tables supplied with the handbook. Samples for tare measurements are also selected by use of random numbers. The number of tare samples for Category A varies as shown in Table 11-1; for Category B, two tares are required for all lot sizes.

Special commodities require special methods of handling, measuring, or calculating. In the case of coffee, for example, the variation of the tare weight of a 1-pound can may be so excessive relative to the product weight that special techniques are required. Additionally, because coffee is vacuum packed after filling, the weight of the air removed must be added to the test weight. Other special commodities are frozen foods, glazed seafood, milk, mayonnaise, and salad dressing.

To summarize the steps recommended by *Handbook 133:*

1. Select category.
2. Define the inspection lot.
3. Choose random samples for testing and tare.
4. Measure each package from Step 3 above.
5. Calculate average, sampling error, and moisture allowance.
6. Evaluate the number of samples exceeding the MAV.
7. Evaluate the comparison of sample average with label claim.
8. Report compliance or noncompliance.

Requirements for compliance are as follows:

1. The number of underweights may not exceed the MAV.
2. The sample average (corrected) must be greater than the label claim.

EVALUATION OF NET CONTENT PERFORMANCE

So far, we have looked at how the government can audit for conformance to net content. Can the same method be used by industry as a quality control line procedure? Not very effectively. The sampling plans for Category B are designed with a 50/50 risk of acceptance/failure for lots whose average is at the labeled weight. This 50% risk may be excessive for the packer who is producing lots complying with the regulation. Category A sampling plans are more discriminating (95% rather than 50%) but provide only after-the-fact control. By the time the required 125 samples from an hour's lot of 10,000 units have been checked, for example, the entire lot may have to be scrapped, whereas the use of statistical quality control charting procedures would highlight a problem for immediate correction.

On the other hand, it is possible (and, in fact, recommended) that available quality control data be periodically statistically analyzed to provide an objective basis for evaluating the costs of overfilling and the risks of noncompliance under current production conditions. Most companies have much of these data readily available. Performance data may be in the form of quality control charts or records (either as daily sheets or summary reports), audit reports, plant performance studies, production volume records, and standard ingredient costs. Specific data such as the following has to be located or assembled:

- Average net weight (or volume)
- Net content variation
- Number of units produced
- Average cost per ounce of ingredients

Regardless of the size of the company or the number of products, the evaluation procedure is essentially the same. If the company is large and there are many products, it would be difficult to conduct a complete analysis at the outset. It would be more advisable to start with one product at one plant, or perhaps a couple of products at two plants, so that comparison between plants could be made at the same time. The following discussion is based on unpublished notes from lectures by C. Kloos at the University of California in April 1988.

In this example, we have four plants producing packages of two varieties of nuts in various sizes. The total number of variables to be compared is seven, a fairly manageable number. These are recorded in columns 1, 2, and 3 in Table 11-3. Start by listing the total number of units produced over the past season (or other period) in column 4. These figures should be readily available from production records if they are not accumulated on quality control records. It may be more difficult to obtain the cost figures to be entered in column 5. If the total costs, labor costs, or overhead costs are difficult to obtain from the production or accounting departments, use the more readily obtainable cost of raw ingredients (and label the column accordingly). If this analysis shows that a savings of 0.5 ounces per can is possible, and if the raw ingredients cost $0.01 per can, then the conclusion would be that a $0.005 savings per can is attainable. If labor, packaging materials, overhead, advertising, and other costs have also been included, there might be considerable discussion as to whether the savings also apply to these factors. Certainly, few would argue the fact that a savings of raw materials as a result of net weight savings is a real benefit. If the costs of manufacturing can be used, combine them with raw materials and enter this figure in column 5.

Next, prepare column 6 by extracting the net weights for the season (or period of study) from quality control records. If running averages are maintained, this should be easy; if not, select data from production records for an arbitrary interval, such as every tenth day, and calculate the average. As a guide, take about 30 day's worth of data. It may include "good" days and "bad" days, but they should not be culled out of the data because, almost by definition, the average net weight is made up of good, bad, and indifferent days. Column 7 data (standard deviations) may be most readily obtained from the range charts ($\bar{R}/d_2$). Although small differences in sigma may not have any appreciable effect on the outcome of this analysis, average sigmas should be used if they have been revised for chart use during the period of this study. By including minor quality control modifications,

Table 11-3. Net Weight Evaluation

(1) Product	(2) Label weight (oz.)	(3) Plant state	(4) Units produced (millions)	(5) Cost ($/oz.)	(6) Avg. wt. (oz.)	(7) Sigma ($\bar{R}/d_2$)	(8) Overfill [(6)−(2)]	(9) Sigmas (Z) overfilled [(8)/(7)]	(10) Cost ($M) [(4) × (5) × (8)]
Walnut	16	CA	420.1	0.03	16.05	0.30	0.05	0.17	630.2
	32	GA	302.0	0.03	32.20	0.35	0.20	0.57	1812.0
	64	NY	56.9	0.03	64.31	0.40	0.31	0.78	529.2
Subtotal			779.0					0.37	2971.4
Pecan	16	CA	650.0	0.04	16.03	0.25	0.03	0.12	780.0
	32	GA	315.5	0.04	32.16	0.29	0.16	0.55	2019.2
	64	NY	106.0	0.04	64.50	0.36	0.50	1.39	2120.0
	128	CA	22.5	0.04	128.90	1.20	0.90	0.75	810.0
Subtotal			1094.0					0.38	5729.2
Total			1873.0					0.38	8700.6

197

this might eliminate distracting arguments during presentation of the study at its conclusion.

Calculating the overfill for column 8 is a simple subtraction of column 2 from column 6. As an example, using the data for 32-ounce pecans packed in New York, the average weight is 32.16 ounces, which is 0.16 ounces over label claim.

Column 9 is the calculated standard deviation from the mean being overfilled, or Z. (You should recall that the value for Z at the average of the distribution equals 0.) This statistic is calculated by dividing the amount of overfill (column 8) by the net-weight standard deviation (column 7). It will be used later in determining the risks of being out of compliance. As an example, take the 16-ounce containers of walnuts produced in California. The formula for Z is

$$Z = \frac{\overline{X} - \mu}{\sigma} = \frac{16.05 - 16}{0.30} = 0.17$$

This means that 16-ounce walnuts in the California plant were overfilled by 0.17 standard deviations. What we have accomplished here is the conversion of overfill in ounces to overfill in standard deviations. This conversion will permit us to calculate the average overfill for the entire company. That, in fact, is what the average figure 0.38 at the bottom of column 9 signifies—the weighted average of overfill for all products is 0.38 standard deviations ($Z = 0.38$). As a matter of interest, most companies overfill by nearly 1.0 standard deviations.

At this point, it might be wise to note that these calculations assume that the net weights are all normally distributed. Although it may not be absolutely accurate to make this assumption, most distributions for filling containers are very nearly normal, with slight skewing on the low side for companies using automatic check-weighers, and slightly on the high side for those who do not. The slight aberrations normally observed for these distributions have only a modest effect on the outcome of the calculations.

The final step in the evaluation of the performance data is the calculation of the overfill costs listed in column 10. The calculation consists of multiplication of the volume produced (column 4) by the average amount of overfill (column 8), times the cost-per-ounce (column 5). For the period selected, the overfill costs for walnuts were $2971.40, and for pecans $5729.20. The total overfill cost was $8700.60.

Having established the costs related to overfilling, we now proceed to the analysis of risk associated with the weight control. The first step is to calculate the percent of individual containers whose net weights are below the stated label weight and enter in column 11. This figure may be found by defining Z as the number of sigmas overfilled (see column 9) and determining the area under the normal curve below $(-Z)$. As an example, the 16-ounce walnut containers produced in California have a Z value of 0.17. By using a table for the areas under the normal curve, it is observed that 43% of the area is below $(-Z)$; that is, 43% of the containers for that particular product are below label weight. This might be more easily seen from Figure 11-1.

Continuing with calculations for column 11, the weighted average percent

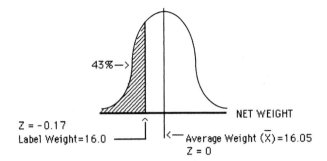

PERCENT BELOW LABEL WEIGHT
16 ounce Walnuts, California

Figure 11-1. Percent below label weight (16-ounces of walnuts, California).

Table 11-4.   MAV for Two Varieties of Nuts

Labeled weight (pounds)	MAV (decimal ounces)
0.94–1.08	0.68
1.88–2.14	1.12
3.90–4.70	2.00
7.90–9.40	3.00

below label weight for all of the items listed amounts to 36%. (Note that the weighted average is obtained by multiplying each value in column 11 by the corresponding number of units produced, as shown in column 4. The total of these calculations is divided by the total number of units produced to obtain the weighted average.)

Handbook 133 has defined unreasonable underfills with the establishment of values for MAV for various classes of product weights. Values for the four weights in our example have been selected from *Handbook 133* and are listed in Table 11-4.

The minimum allowable weight is determined by subtracting the appropriate MAV from the label weight. For example, the 16-ounce package is between 0.94 and 1.08, and has a MAV of 0.68. By subtracting 0.68 from 16, the minimum value of 15.32 results. This is entered in column 12 of Table 11-5.

Column 13 lists the percent of individual containers that are below the minimum allowable net weight. The calculations are similar to those used to determine column 11, in which Z determines the area under the normal curve:

Table 11-5. Risk Analysis

(1) Product	(2) Label weight (oz.)	(3) Plant state	(11) % Below label (avg. %)	(12) Mini-mum (oz.)	(13) Below MAV (%)	(14) Plan A $P(C>1)$ (%)	(15) Plan B $P(C>0)$ (%)	(16) Plan A limit	(17) Plan A ($n=30$)	(18) Plan B ($n=10$)
Walnut	16	CA	43	15.32	0.7	1.9	6.8	15.89	0.2	29.9
	32	GA	28	30.88	0.008	0.0	0.08	31.87	0.0	3.5
	64	NY	22	62.00	0.0	0.0	0.0	63.85	0.0	0.7
Subtotal			36		0.38	1.0	3.7		0.1	17.5
Pecan	16	CA	45	15.32	0.2	0.2	2.0	15.91	0.4	35.2
	32	GA	29	30.88	0.0	0.0	0.0	31.89	0.0	4.1
	64	NY	8	62.00	0.0	0.0	0.0	63.87	0.0	0.0
	128	CA	23	125.00	0.12	0.01	0.6	127.56	0.0	0.9
Subtotal			36		0.12	0.1	1.2		0.2	22.1
Total			36		0.23	0.5	2.2		0.2	20.2

$$Z = \frac{\text{Minimum allowable net wt.} \left(\text{col. 12}\right) - \text{Avg. net wt.} \left(\text{col. 6}\right)}{\sigma\left(\text{col. 7}\right)}$$

As an example, the first entry in column 13 is calculated for 16-ounce walnuts produced at the California plant. The individual containers could be as low as 15.32 ounces before being declared below minimum. The percentage of such containers may be found by first solving for Z in the equation above:

$$Z = \frac{15.32 - 16.05}{0.30} = -2.43$$

The area under the normal curve is found in the Z tables and is shown to be 0.7%. By counting these calculations for all of the products and taking their weighted averages, it is shown that the company has produced, on average, 0.23% of its containers below the allowable minimum.

The next calculations will determine the probability of failing to meet the compliance criteria under Plan A and Plan B of *Handbook 133*. Under Plan A, of the 30 samples selected, no more than one container may fall below the MAV, and the sample average must be no less than two standard deviations of the mean below the label weight. Under Plan B, of the 10 samples selected, no samples may be below the allowable minimum, and the sample average must be equal to or greater than the label declaration. A sample calculation for Plan A and for Plan B follows.

A defect is defined as a container with net weight below the allowable minimum. The probability of finding two or more defects in a sample of 30 is expressed as $\Pr(C > 1, n = 30)$ and will be entered in column 14. Similarly, the probability of finding one or more defects in a sample of 10 is expressed as $\Pr(C > 0, n = 10)$ and will be entered in column 15.

These two calculations represent the risk of being out of compliance on the basis of individual containers in the test sample. They are based on the binomial probability distribution where p (the percent defective) is that which is shown in column 13 for each product. As an example, refer to the first item in column 13, where $p = 0.7\%$, or 0.007. These values are then inserted into the binomial formula:

$$\Pr(C > 1, n = 30) = 1 - [\Pr(C = 0) + \Pr(C = 1)]$$

$$= 1 - [(.993)^{30} + 30(.007)^{1}(.993)^{29}]$$

$$= 1 - .981$$

$$= .019 \text{ or } 1.9\%$$

This is the first entry in column 14 for 16-ounce walnuts produced in California.

$$\Pr(C > 0, n = 10) = 1 - \Pr(C = 0)$$

$$= 1 - (.993)^{10}$$

$$= 1 - .932$$

$$= .068 \text{ or } 6.8\%$$

This is the first entry in column 15 for 16-ounce walnuts produced in California.

Continuing for each of the 7 products, we find that the company has a 0.5% chance of finding 2 or more defects in a sample of 30 (Plan A) and a 2.2% chance of finding 1 or more defects in a sample of 10 (Plan B).

Now that the risks of exceeding the maximum allowable variation have been determined, the next step is to find the risks associated with failing to reach an acceptable average weight. Under Plan A, the sample average can be no lower than two standard deviations of the mean $s_{\bar{X}}$ below the label weight. To determine the allowable limit under Plan A, first determine the value of $s_{\bar{X}}$. This is found from the equation

$$s_{\bar{X}} = \frac{\sigma}{\sqrt{n}}$$

where s is sigma for the product (see column 7) and n is the sample size. For 16-ounce walnuts from California, we find

$$s_{\bar{X}} = \frac{0.30}{\sqrt{30}} = 0.055$$

The allowable limit (column 16) is found by subtracting $2s_{\bar{X}}$ from the label weight. For 16-ounce California walnuts,

$$\text{Allowable limit} = 16.0 - 2(0.055) = 15.89$$

Similarly, the remaining product allowable limits are calculated and entered in column 16. This data may then be used to complete the calculation for column 17, the risk of finding a mean below the limit for Plan A, and column 18, the risk of finding a mean below the limit for Plan B.

For Plan A, calculate the number of standard deviations of the mean $s_{\bar{X}}$ being overfilled. This is done by using the equation

$$Z_{\bar{X}} = \frac{\left[\text{Limit} \left(\text{col. } 16\right) - \text{Avg. net wt.} \left(\text{col. } 6\right)\right]}{\sigma \left(\text{col. } 7\right) / \sqrt{n}}$$

The risk of finding a mean below the limit (column 17) is then defined as the area below $Z_{\bar{X}}$ under the normal curve. For 16-ounce packages of walnuts from California,

$$Z_{\bar{X}} = \frac{15.89 - 16.05}{0.30/\sqrt{n}} = -2.92$$

Therefore, the probability that a sample mean of 30 containers of this product is the area below $Z = -2.92$. By using the table for areas under the normal curve, we find the area corresponding to $Z = -2.92$ is 0.002, or 0.2%. The remaining products are similarly evaluated and weighted averages calculated to give an overall risk of company noncompliance of 0.2%.

The final calculations for this analysis are for Plan B compliance for the averages, column 18. Under this plan, the sample average must be at or above the label weight. Therefore, the limit for the sample average is the same as the label weight. Calculate the number of standard deviations of the mean $s_{\bar{X}}$ being overfilled above the declared label weight. Define $s_{\bar{X}}$ by using the equation

$$s_{\bar{X}} = \frac{\sigma}{\sqrt{n}}$$

$$= \frac{\sigma\left(\text{col. } 7\right)}{\sqrt{10}}$$

The number of standard deviations of the mean being overfilled is found using the equation

$$Z_{\bar{X}} = \frac{\text{Label wt.} - \text{Avg. net wt.} \left(\text{col. } 6\right)}{\sigma\left(\text{col. } 7\right)/\sqrt{10}}$$

$$= \frac{16.0 - 16.05}{0.30/\sqrt{10}} = -0.53$$

The probability that a sample mean of 10 packages of this product is therefore the area below $Z = -0.53$ under the normal curve, which is found in the tables to be 0.299, or 29.9%. When the remainder of the product data are calculated and the weighted averages determined, the overall company risk of noncompliance with Plan B requirements is 20.2%.

PROCEDURES FOR SETTING FILL TARGETS

The calculations in Tables 11-3 and 11-5 are based on hypothetical quality control data to reveal

- The cost of overfilling (column 10).
- The chance of being out of compliance with the existing quality control systems for net weight.

- (under Plan A) Failure to meet MAV requirements (column 14) and failure to meet sample average weight (column 17).
- (under Plan B) Failure to meet MAV requirements (column 15) and failure to meet sample average weight (column 16).

As vital as this information is to top management, it still leaves three questions:

1. Are these levels of cost and risk satisfactory?
2. At what levels should the company be operating?
3. Can the company meet these new levels?

It is unlikely that a quality manager would have the gall to attempt to answer the first question. In the first place, top management would like to see both cost and risk levels at zero, and probably realizes that both goals are unattainable. Only management is in a position to answer this question, and they probably will not. Instead, they will probably ask, "Can't we do better?"

The second question can be calculated and will be demonstrated shortly. This brings us to the third question. Because the answer depends on the value of the standard deviation for each of the packaging lines, it can be determined by conducting a series of production floor tests to establish the effect of production variables (line speed, product uniformity, density, humidity, equipment condition, modified equipment, and so on) on the standard deviation of the weights attained. Because performing these tests is somewhat costly, it would be well to first calculate the theoretical goals, and then check if the lines are capable of operating under those conditions. If not, then tests should be designed to explore methods of meeting the calculated standards.

A procedure for calculating the target net weight using the data for 16-ounce walnuts produced in California (the first line of Table 11-4) follows. The data required for this example are

Label weight (from column 2) = 16 avg. ounces
Sigma (from column 7) = 0.30 avg. ounce
Target 5 below label weight (from management) = 25%

Assuming that the weights are distributed normally, the target value (μ) is set such that 25% of the area under the normal curve is below the label (16.0 ounces). By referring to Table A-2 in the Appendix, it is shown that Z (the number of standard deviations) is equal to -0.675. Solving for μ in the equation

$$Z = \frac{\text{Label weight} - \mu}{\sigma}$$

$$-0.675 = \frac{16.0 - \mu}{0.30}$$

$$\mu = 16.20$$

Therefore, a target net weight as low as 16.20 ounces will result in no more than 25% below the label weight, as illustrated in Figure 11.2.

Selecting a 25% maximum is one way to set target weights, but it is purely arbitrary. Another way is to select a risk of failing to comply with *Handbook 133* requirements. This, too, is arbitrary in that the risk is a management decision, but it is more meaningful.

As stated earlier, *Handbook 133* has two requirements for compliance: (1) the MAV and (2) the average net weight. The first calculation will be to determine the target net weight needed to assure a minimum risk of producing an unacceptable number of short-filled containers below the MAV. Defining *minimum risk* is a management decision, but should management be unwilling to provide a figure, some reasonable low risk must be assumed in order to proceed. This is not as difficult as it might seem. The figure 98% compliance is occasionally heard in relation to risk. This implies that, of the samples selected for net content evaluation by the authorities, no more than 2% will be not in compliance. From a practical standpoint, if an individual food processor's products were sampled 10 times a year, it would take 10 years to accumulate 100 samplings; no more than two of those examinations might be out of compliance. It is not unreasonable to assume that, with a 2% risk, a processor might have no violations for 10 times that period, or 100 years. A company might not be selected for any sampling for several years. On the other hand, it is only fair to surmise that once a violation has been observed by the authorities, a company's products will come under close scrutiny for some time.

For this example, management is willing to take a 2% risk of failing the compliance requirement for individual containers. Under Plan A, compliance requires, at most, one defect (container below the MAV) in a sample of 30. This is shown as

$$\text{Risk} = \Pr(C > 1, n = 30) = 1 - [\Pr(C = 0) + \Pr(C = 1)]$$

$$0.02 = 1 - [(1 - p)^{30} + 30(1 - p)^{29} (p)]$$

Solving for p (percent defective), it is found that $p = 0.007$ or 0.7%. To solve for μ, such that only 0.7% fall below the MAV (15.32 ounces), the appropriate Z

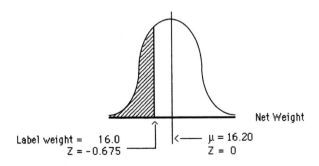

Figure 11-2.   Selection of target weight.

value must first be found, using Table A-2 in the Appendix. Having found $Z = 2.455$, solve for μ using the equation

$$Z = \frac{MAV - \mu}{\sigma}$$

$$-2.455 = \frac{15.32 - \mu}{0.30}$$

$$\mu = 16.06$$

This means that if the target is set at 16.06, there is a 2% risk of noncompliance for individual underfills under Plan A of *Handbook 133*.

Similar calculations are performed for Plan B.

$$Risk = Pr(C > 0, n = 10) = 1 - Pr(C = 0)$$

$$0.02 = 1 - (1 - p)^{10}$$

It might be more suitable to use logarithms to solve this equation:

$$(1 - p)^{10} = 0.98$$

$$10 \log(1 - p) = \log(0.98)$$

$$\log(1 - p) = 0.00202$$

Solving for antilogarithms,

$$1 - p = 0.9980$$

$$p = 0.002, \text{ or } 0.2\%$$

The appropriate Z value for 0.2% below the MAV is -2.88. Now the calculation for μ is completed by using the equation

$$-2.88 = \frac{15.32 - \mu}{0.30}$$

$$\mu = 16.18 \text{ ounces.}$$

This is the target to assure 2% compliance risk for underfills under Plan B of *Handbook 133*.

Now that the calculations for MAV compliance have been calculated, the next step is to calculate compliance for average weight. The averages are normally distributed, with standard deviations defined as

$$S_{\bar{X}} = \frac{\sigma}{\sqrt{n}}$$

The target μ is set so that the area under the normal curve below the minimum for the average is equal to the desired level of risk. As above, a 2% risk will be used. Under Plan A, $n = 30$, and the minimum sample average must be less than two standard deviations:

$$\text{Minimum} = \text{Label weight} - 2S_{\bar{X}}$$

$$= 16.0 - 2\left(\frac{0.30}{\sqrt{30}}\right)$$

$$= 16.0 - 0.11$$

$$= 15.89$$

The appropriate Z value ($Z_{\bar{X}}$ for a 2% risk is –2.054. Therefore, the target net weight (μ) can be found by solving the equation

$$-2.054 = \frac{15.89 - \mu}{0.30/\sqrt{30}}$$

Hence, $\mu = 16.00$ ounces (see Fig. 11-3).

Under Plan B of *Handbook 133*, the sample size is usually 10; therefore, $s_{\bar{X}} = \sigma/\sqrt{10}$. Note that there are no allowances for the sample mean to be below the label weight under Plan B. The desired target weight for a 2% risk can be found by solving the equation

$$-2.054 = \frac{16.00 - \mu}{0.30/\sqrt{30}}$$

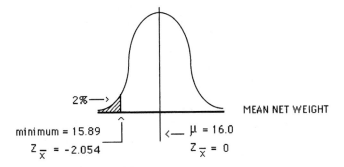

Figure 11-3. Compliance under Plan A.

Hence, $\mu = 16.11$ ounces (see Fig. 11-4).

At this point, it might be wise to tabulate the results of all of the preceding target weight calculations. A comparison of the two compliance programs is shown in Table 11-6.

From this table, management should be in a position to select one program. The final decision may be based on an analysis of the cost/risk trade-off. To derive the relationship between the cost of overfilling and the risk of noncompliance, a company should determine the overall cost for 1-sigma overfill. This is useful because an overfill policy is usually expressed in terms of a certain number of sigmas overfill and provides for the evaluation of alternative overfill policies without considering each product separately. Again, we shall use the 16-ounce package of walnuts processed in the California plant. The data on which the calculations are based are as follows:

Label weight = 16.0 ounces
Average net weight = 16.05 ounces
Sigma (σ) = 0.30 ounces
Volume = 420,100,000 units produced
Cost = $0.03 per ounce

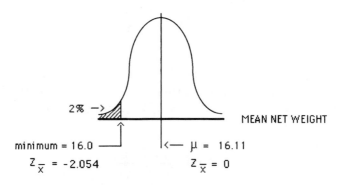

Figure 11-4.   Compliance under Plan B.

Table 11-6.   Target Weight Matrix

Criteria	Plan A	Plan B
25% below the label weight	16.20 oz.	16.20 oz.
2% risk of noncompliance		
Individual underfills	16.06 oz.	16.18 oz.
Sample average	16.00 oz.	16.11 oz.

Note: Target net weights are for 16 ounces of walnuts from California (to comply with *Handbook 133*).

The cost of 1 sigma is found by determining the total cost of overfills and dividing that number by the level of sigma overfill:

420,100,000 units $\times$ 0.05 ounces $\times$ \$0.03 per ounce = \$630,150

$$\frac{\text{Amount overfilled}}{\sigma} = \frac{0.05}{0.30} = 0.17$$

$$\text{The cost of overfilling} = \frac{\$630,150}{0.17} = \$3.71 \text{ million}$$

Similar calculations for the cost for overfilling all products will provide a rational means of selecting a cost/risk policy.

The preceding calculations are based on a system for complying with the requirements of *Handbook 133* and selecting a risk and cost level for each product. There are other successful techniques, but to be assured that the company is in compliance, it is advisable to periodically run through these calculations. Some companies believe that mere compliance with the law is not sufficient for their product lines, and it is conceivable that filling techniques could be improved to the point where weights are well within the compliance limits. In general, this may be accomplished by progressively reducing the sigma value for each production line through experimentation, training, maintenance, more uniform product, and improved equipment. Simply comparing the control limits of $\overline{X}$ and R charts as these improvements are introduced will indicate the direction for tighter weight control. This technique can cut the value of sigma to the point where the target value of a 16-ounce container can be a scant 0.016 ounces above target weight with no loss in productivity. The reduction of associated costs related to product giveaway on a high-speed line are often spectacular.

12   Design of Experiments

Design of experiments consists of a series of techniques used to explore various operating conditions of a process with the goal of finding lower raw material, process, and product costs, improved or new product performance, less product variability, or higher production capacity to meet increasing demand. In large companies with specialized departments, experimental design is the province of engineers, food scientists, or research and development, but quality control personnel may become involved with performing, and sometimes evaluating, industrial experimentation. Consequently, an understanding of the techniques of this discipline is valuable. In smaller companies, where the task of industrial experimentation often falls to the quality control manager, a knowledge of statistical design of experiments is imperative.

Perhaps we should consider again the meaning of the word *quality*. Among others, the clothing industry commonly uses the word to signify superior appearance or performance. Throughout this book, we have referred to quality as a *level*, not necessarily a superior level or the highest level. Many classes of products need not be of high quality to perform their functions; in fact, high quality may sometimes be a deterrent to marketability. Disposable tools for assembling toys or furniture may be made of a stamped steel sheet rather than heavy-duty, forged, chrome steel alloys. Pie-grade fruit to be further processed into pies, jellies, or fillings need not consist of unbroken, uniform pieces. Salt used to melt sidewalk ice would gain nothing but cost if manufactured to the same purity and quality levels as food-grade salt. Plastic film packaging certainly does not provide the superior protective quality of hermetically sealed metal containers, but many foods with short shelf life or an appearance which could be advantageously displayed in clear plastic need not—and perhaps should not—be sealed in more expensive containers. Does statistical design of experiments have any place in industries such as these? Of course it does. It may be possible to improve quality level at no additional cost, thus providing a marketing advantage even to throwaway products. It may also be possible to design experiments which lead to maintaining the quality level with higher rates of production, less expensive or alternative raw materials, or less labor.

At the other end of the quality scale, a series of experiments could be designed to eliminate defects in costly food products, such as discoloration in blanched almonds used for stuffing prunes. New process methods may be explored to improve the color retention of freeze-dried vegetable products; or modification of cooking and packaging methods might improve the sensory characteristics and extend the life of shelf-stable meals. The list is endless, and opportunities to increase profits while improving the quality level are unlimited.

Occasionally, the need for an experimental effort is thrust upon a processor. Frequently, the driving force behind industrial experimentation is the threat of a changing market: consumers' tastes change, or a competitor modifies his product or introduces a new one. Sometimes, a raw material source dries up or becomes more costly due to weather, an increased demand from other markets, or other conditions. Disparaging press reviews of suspected environmental damage caused by a product or its packaging often force a company into a crash program of experimentation.

The main goal of a statistically designed experiment is to obtain unambiguous answers to specific questions. Modifying a production facility or running a non-standard experimental process is expensive. There is lost production, wasted wages and overhead, and the likelihood of creating unusable scrap material. It is important that an experiment be adequately planned to provide specific answers and to avoid having to repeat the procedure to correct mistakes. Consider a test casually run with a modified formula, after which the line foreman discovers that the test was produced on the wrong machine at a different temperature from normal production. Even if the results of the test showed an improved product, the foreman is left with doubts about the reasons for success: Was it the formula? the machine? the temperature? There is no room for careless planning or execution in a statistically designed experiment.

There is another reason for careful planning: It is rarely possible to introduce a single variable in an experiment. For example, suppose a test is suggested which will result in a lower moisture product by increasing dryer airflow. The increased airflow might cause shorter product retention time in the dryer, and to compensate for the resulting decrease in BTU transfer to the product, the temperature would also have to be raised. It is also rarely possible to introduce variables in a process which will deliver a single change in the final product. In the example involving increased dryer airflow, it is likely that not only will the moisture be affected, but product color will probably be darker because of the added temperature, and product density might be appreciably altered. These new conditions may or may not be undesirable, but careful planning before running the experiment should eliminate surprises in the outcome and reduce the necessity of conducting additional experiments because of unanswered questions. The results of an experiment to show the efficiency of using large pineapples as a raw material would indicate that major labor savings could result from eliminating many of the trimmers and replacing them with a large-fruit mechanical trimmer. But this experiment ignores the real-world fact that all pineapples are not large; trimmers will still be required for the normal-crop-size raw materials received during the canning season.

The two goals of statistical design of experiments (unambiguous answers and low-cost testing) are reached through the elimination of extraneous variables and the ability to handle many factors simultaneously. Extraneous variables commonly encountered are the day-to-day differences found in any production facility: different employees, varying motivation of the workers, environmental changes (temperature, humidity, light, ventilation, vibration, odor), day of the week, proximity to holidays or payday, and even local news events. For the most part, these variables are related to people. There are also variables related to machines that appear to be identical but that actually have unique operating characteristics. When the output of two filling machines is compared, they are rarely found to produce identical weight dispersions. This may be the result of machine speed differences, wear, product flow, or machine location. Raw materials used in a process may vary from supplier to supplier and hour to hour. In some industries, this variability may be controllable: A hopper car of plastic pellets used for molding may not vary in composition from top to bottom, but a truckload of tomatoes may have drastic quality differences due to causes as simple as location in the pile—crushed on the bottom, sunburned on the top, and cold in the front. There are also field-to-field chemical variations and maturity differences. Finally, there are often uncontrolled differences between the technicians conducting the tests and the evaluators interpreting the results.

Often there are conditions under which it is not possible to either eliminate these variables or hold them constant. There are also circumstances under which it would be undesirable to attempt to hold them constant. For example, suppose a test is under consideration to establish the ruggedness of a process when external and uncontrollable variables are commonly present. Such a test might be designed to determine which raw ingredients are resistant to uncontrollable process fluctuations and can be depended on to result in a uniform quality product. Therefore, it is important to avoid confusing *essential external variables* with *experimental variables*. Uncontrolled variables may often be handled by randomly selecting them for inclusion in the test design.

There are three objectives in designing experiments:

1. All design factors and response variables associated with the problem being investigated should be included.
2. Design factors should be appropriately balanced so that none are either overemphasized or overlooked.
3. The reproducibility of results should be both precise and accurate.

Precision refers to the variability or uniformity of results and is generally characterized by the standard deviation σ. *Accuracy* is a measure of the difference between the measured and the theoretical results and is expressed as $(\bar{X} - \mu)$. For example, a filler scale might produce highly precise weights—all containers within 0.001 ounce—but the accuracy might be unacceptable at 0.75 ounces overweight.

The number of variables in a design depends on the complexity of the investigation, the production time available, and the cooperation of production and other managerial personnel. Tests designed to explore a single variable are usually easiest to sell to management, because they are readily understood and appear to be the least costly. Some limitations have already been discussed. Experiments may be conducted *off-line,* which generally refers to experiments conducted in the pilot plant, in the laboratory, on commercial equipment at an equipment manufacturer's facility, or on plant equipment when the plant is not in normal production. Off-line experiments are intended to eliminate interference with production schedules, but they may not necessarily be the most cost-effective.

In addition to experiments with only a single variable, testing with slight changes in a few small operating conditions at a time is frequently a successful way to design experiments. Small changes will probably not affect production output quantity or quality appreciably but will show the direction for further improvement. Procedures of this type have been formalized as *evolutionary operations* (EVOP). A third general experimental technique examines several different operating conditions, usually over a relatively short period of time.

ELIMINATION OF EXTRANEOUS VARIABLES

Example
Which of four recipes makes the best cherry pie—A, B, C, or D? In order to evaluate the products, a sensory testing panel requires four pies of each recipe. The laboratory is unable to bake more than four pies in a day. The most obvious solution would be to bake one pie of each type (A, B, C, and D) on one day, and submit them to the testing panel each day for four days. The problem with this setup is that it does not take into account day-to-day variations that may occur among the judges or changes in the laboratory personnel making the pies. These are *extraneous variables.* The conclusions from such a test might improperly disclose apparent differences between pies which are truly differences between days.

A method which minimizes these day-to-day differences is known as a *randomized block design.* Under this plan, different recipes are made up each day, but not necessarily one of each type. However, as shown in the table below, note that each pie has been baked four times over the four-day period.

Day 1	Day 2	Day 3	Day 4
B	A	C	C
D	C	A	D
A	B	D	A
C	D	B	B

In a randomized block design, all treatments are assigned to a block, which is defined as a homogeneous set of conditions for comparing the treatments. In the pie example, days are blocks. Again, the main purpose of this plan is to avoid confounding of extraneous variables with primary variables. A second purpose is to improve precision by removing experimental error from effect.

Another example of a randomized block design is one in which four bread formulas are compared for loaf volume when baked in a microwave oven. The primary variable here is the bread formula; the extraneous variable is the type of oven used. The formula data and oven data are referred to as *design variables* or *design factors* (variables over which the designer has control); the loaf volumes are referred to as the *response variables* (or *outcome variables*).

The first step in designing this test is to develop a protocept. This is a laboratory benchtop formulation satisfying the general requirements for the product to be tested. Some of the considerations might be size, weight, texture, flavor, or color. Having established the attributes of the product, the protocept is then ready to be subjected to a series of tests, the first of which should be designed as simply as possible.

Note the outcomes in the following table:

		Ovens		
Formula	GE	TA	SA	AM
1	156	143	160	137
2	150	151	157	129
3	112	137	126	109
4	162	151	175	149

It appears that the smallest-volume loaves are produced by Formula 3. Regardless of which oven is used, this formula produces the lowest numbers for volume. Formula 4 produces higher volumes than any other formula for a given oven, but note that the volume in the TA oven (151) is smaller than the volume in the GE oven (156). This is extraneous information. The goal was to compare formulas, not microwave ovens. This is a one-factor design: bread formula.

An alternative plan for the earlier pie-baking problem is called the *Latin square*. It is designed to control for two blocking factors: day and sequence. You will note in the Latin square following this paragraph that each pie is baked only once on each day and that the first pie baked is different each day. This eliminates the possibility of drawing incorrect conclusions if the first pie baked is always different from subsequent pies due to lack of oven prewarming. The two extraneous variables here are (1) the day the pies are baked and (2) the sequence of baking.

Sequence	Day 1	Day 2	Day 3	Day 4
1	A	D	C	B
2	B	A	D	C
3	C	B	A	D
4	D	C	B	A

When a large number of variables must be compared at the same time, a modification of the randomized block design may be used. Assume that we wish to compare seven pie recipes. The sensory panel judges can evaluate no more than three pies before losing their ability to detect flavor differences. The design which satisfies these requirements is called an *incomplete block design*. The seven recipes for pies are shown as A through G below in the incomplete block following this paragraph. Note that each judge evaluates only three pies and that each pie is sampled by three different judges. The design is carefully balanced so that each group appears the same number of times and each formula appears the same number of times. Note also that the pair AE occurs only once, as do the pairs AC, EC, and so on; that is, each pair of recipes appears only once. In this manner, each pie recipe can be compared independently of judge-to-judge variation. Prepared tables of incomplete block designs are available in the literature.

			Judges			
1	2	3	4	5	6	7
A	G	A	C	B	C	D
E	F	B	G	E	D	G
C	A	D	B	F	F	E

Table 12-1 illustrates several commonly used examples of incomplete block designs.

Statisticians have developed many types of block designs to handle experiments other than the simple ones shown here. The most commonly used block designs include:

- Randomized block
- Balanced incomplete block
- Latin square
- Graeco-Latin square (for three-block variables)
- Hyper-Latin square (for more than three variables)
- Incomplete Latin square

HANDLING MANY FACTORS SIMULTANEOUSLY

Looking at the cherry pie formulation from another viewpoint, suppose we are interested in evaluating the effect of three factors:

1. Butter (B) versus margarine (M)
2. Cherry supplier A or B
3. Amount of sugar—high or low

Table 12-1.   Incomplete Block Designs

4 treatments, 2 per block			6 treatments, 2 per block		
Block	Treatments		Block	Treatments	
1	a	b	1	a	b
2	c	d	2	c	d
3	a	c	3	e	f
4	b	d	4	a	c
5	a	d	5	b	e
6	b	c	6	d	f
			7	a	d
			8	b	f
			9	c	e
			10	a	e
			11	b	d
			12	c	f
			13	a	f
			14	b	c
			15	d	e

5 treatments, 2 per block			6 treatments, 3 per block			
Block	Treatments		Block	Treatments		
1	a	b	1	a	b	e
2	c	d	2	a	b	f
3	b	e	3	a	c	d
4	a	c	4	a	c	f
5	d	e	5	a	d	e
6	a	d	6	b	c	d
7	b	c	7	b	c	e
8	c	e	8	b	d	f
9	a	e	9	c	e	f
10	b	d	10	d	e	f

5 treatments, 3 per block				7 treatments, 3 per block			
Block	Treatments			Block	Treatments		
1	a	b	c	1	a	b	d
2	a	b	e	2	b	c	e
3	a	d	e	3	c	d	f
4	b	c	d	4	d	e	g

Table 12-1. (*continued*)

5 treatments, 3 per block				7 treatments, 3 per block			
Block	Treatments			Block	Treatments		
5	c	d	e	5	a	e	f
6	a	b	d	6	b	f	g
7	a	c	d	7	a	c	g
8	a	c	e				
9	b	c	e				
10	b	d	e				

The number of pies that can be baked is restricted to four. A typical nonstatistical method might involve changing one variable at a time. The first pie would contain butter, cherries from supplier A, and low sugar. The second pie would contain margarine, but otherwise be the same as the first pie. The third pie would be the same as the first but with cherries from supplier B. The fourth pie would be the same as the first but with a high level of sugar. This nonstatistical arrangement is as follows:

Shortening	Cherries	Sugar
B	A	LO
M	A	LO
B	B	LO
B	A	HI

This plan adequately compares some of the pie variables with the first pie but does not cover all of the interactions. For example, there is no formula to compare the first pie with one made with margarine and high sugar; nor is there a pie with margarine and cherries from supplier B. To include all combinations would require many more pies than could be baked at one time, but with the four combinations above, the three variables are not treated uniformly. Only one test uses margarine, and three use butter. Only one test has high sugar, and three have low.

By using a statistical plan, some of these deficiencies may be overcome. In the following statistically designed plan, each variable is examined the same number of times: two butters and two margarines, two high sugars and two low sugars, two supplier A and two supplier B. Note also that each variable is tested one time against each of the other variables. For example, butter (B) is tested with supplier A and supplier B, and with high and low sugar. Examining the following statistical plan one finds that all variables are similarly compared once with each other:

Shortening	Cherries	Sugar
B	A	LO
B	B	HI
M	A	HI
M	B	LO

One example of the advantage of the statistical plan over the nonstatistical plan is the fact that the statistical plan will compare the effect of butter over margarine for each supplier and at each level of sugar. The nonstatistical plan does not compare the effect of butter over margarine either for supplier B or for high sugar. Another advantage of the statistical plan is that all formulas are used in estimating the effects. The shortening effect is the difference between the average of the first two formulas and the average of the last two formulas. By being able to use averages, the standard error is reduced by a factor of $\sqrt{2}$. This small standard error enables the statistical plan to detect small differences.

Multifactor experimental plans, which handle many factors simultaneously, include

- Full factorial designs
- Fractional factorial designs
- Response surface designs
- Mixture designs

Full Factorial Designs

A full factorial design includes all possible combinations of all of the conditions. In the case of the cherry pie tests, if there had been no restrictions on the number of pies baked at one time, a full factorial design might have been used and would appear as follows:

Shortening	Cherries	Sugar
B	A	HI
B	A	LO
B	B	HI
B	B	LO
M	A	HI
M	A	LO
M	B	HI
M	B	LO

The simplest factorial design is called a *2 × 2 design*. It considers the responses of a test in which two factors (variables) are examined at two levels (values). In the following example, two variables (temperature and concentration) will be studied at two levels (low and high). The response to be evaluated shall be the process yield. Figure 12-1 is a diagrammatic representation of the experiment.

A total of four conditions will include all possibilities for conducting this test design (see Table 12-2).

The experiment is subsequently conducted under these four sets of conditions, yielding the results shown in Figure 12-2 and Table 12-3.

Figure 12-1. A 2 × 2 design.

Table 12-2. 2 × 2 Factorial Experimental Design Conditions

Condition number	Concentration (%)	Temperature (°C)
1	13.5	124
2	14.5	124
3	13.5	128
4	14.5	128

Figure 12-2. Data from a 2 × 2 experiment.

Table 12-3. 2 × 2 Factorial Experimental Design Data

Condition number	Concentration (%)	Temperature (°C)	Yield
1	13.5	124	60.2
2	14.5	124	67.6
3	13.5	128	73.2
4	14.5	128	76.2

Examining the effects of increasing the concentration from 13.5 to 14.5% gives

Concentration Effect

$$67.6 - 60.2 = 7.4 \text{ @ Temperature } 124°C$$
$$\underline{76.2 - 73.2 = 3.0} \text{ @ Temperature } 128°C$$
$$71.9 - 66.7 = 5.2 \quad \text{Average yield}$$

The yield increases at either temperature, but to a greater extent at 124°C. If the process were operated at the higher concentration, we might expect that the lower the temperature (between 124°C and 128°C), the better the yield:

Temperature Effect

$$73.2 - 60.2 = 13.0 \text{ @ Concentration } 13.5\%$$
$$\underline{76.2 - 67.6 = 8.6} \text{ @ Concentration } 14.5\%$$
$$74.7 - 63.9 = 10.8 \quad \text{Average yield}$$

Similarly, if we increase the temperature while holding the concentration at 13.5%, the yield increases by 13.0%. At the high concentration, increasing the temperature increases the yield by 8.6%. From this simple design, a surprisingly large amount of useful information emerges.

The interaction effect can be quantified as well. The difference between the concentration effect from high to low temperature is 7.4 – 3.0 = 4.4. The difference between the temperature effect from high to low concentration is 13.0 – 8.6 = 4.4. One might expect to measure the interaction as 4.4, but the definition used by statisticians is based on half this difference and is expressed as 2.2 ± 2.2. To explain the interaction effect, consider this fact: If there were no interaction effect, the main effects of both variables would be the same at both levels of the other variable; that is,

Concentration effect = 5.2 @ Temp. = 124°C and 128°C

Temperature effect = 10.8 @ Conc. = 13.5% and 14.5%

If we now consider the actual effects at the high level of the other factor,

$$\text{Concentration effect} = 3.0 \ @ \ \text{Temp.} = 128°C$$
$$\text{Temperature effect} = 8.6 \ @ \ \text{Conc.} = 14.5\%$$

resulting in

$$\text{Interaction effect} = 3.0 - 5.2 = -2.2 \quad \text{and} \quad 8.6 - 10.8 = -2.2$$

Figure 12-3 answers the question, "Does it matter what the temperature level is when the concentration changes are examined?" It is obvious that there is a large difference in yield when moving from 13.5% to 14.5% concentration on the 124°C temperature graph, and there is a smaller difference in yield when moving from 13.5% to 14.5% concentration on the 128°C temperature graph. By changing the graphed axes to yield and temperature, the effect is again observed: There is a larger effect at 124°C than at 128°C.

The yield values 60.2, 67.6, 73.2, and 76.2 in Figure 12-4 were taken from an actual test, and although only a single value was shown in order to simplify the principle, actually two sets of values were generated from duplicate tests. A summary of the preceding discussion of data, calculations, interactions, and main effects follows:

Main effect A = 5.2   Main effect B = 10.8   Interaction effect AB = −2.2

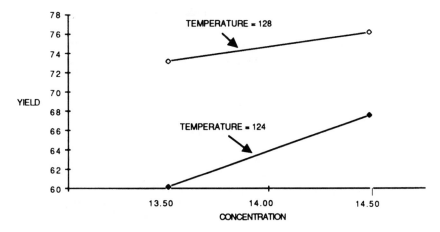

Figure 12-3.   Interaction plot of concentration by temperature.

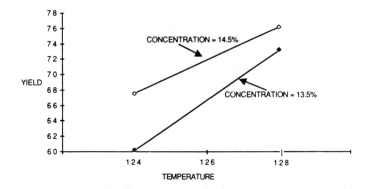

Figure 12-4.   Interaction plot of temperature by concentration.

Condition number	Replication 1	Replication 2	Average	Standard deviation	Variance
1	59.1	61.3	60.2	1.56	2.42
2	68.2	67.0	67.6	0.85	0.72
3	72.4	74.0	73.2	1.13	1.28
4	74.9	77.5	76.2	1.84	3.38
				Total	7.80

Average variance = 7.80/4 = 1.95
Average standard deviation = 1.396 ~ 1.40
Standard error of the effect = 1.4 / $\sqrt{2}$ = 1.0
Main effect A is 5.2 ± 2   (3.2 to 7.2)
            B is 10.8 ± 2   (8.8 to 12.8)
Interaction effect AB is –2.2 ± 2   (–0.2 to –4.2)

An example of a full factorial design with statistical analysis is the examination of the effect of a fungicide and two pesticides on the egg production of pheasants. Four observations are available for each combination.

Pheasant Egg Counts

	Pesticide Levels					
	None		Level A		Level B	
No fungicide	15	29	7	15	17	19
	18	30	9	18	16	21
With fungicide	12	15	13	10	9	11
	10	9	6	8	14	13

An analysis of variance shows that there is an effect from both the fungicide and from the pesticide.

Analysis of Variance

Source	df	MS	F	
Fungicide	1	294.00	16.04	$(p < .01)$
Pesticide	2	87.17	4.75	$(p < .05)$
$F \times P$	2	36.50	1.99	
Error	18	18.33		

Full factorial designs are used if there are only two or three factors to explore. They are also useful with higher-order interactions; that is, some factors have little effect on outcomes of an experiment until a threshold is reached, after which the effect may be significant. The statistician may be unaware of these interactions and must depend on the experience of quality control or line personnel to provide information on their importance so that they may be incorporated into the experimental design. Another application requiring full factorial design is the analysis of qualitative independent variables, such as the number of different suppliers of raw materials. If there are four suppliers available, all four must be included in the test plans.

Fractional Factorial Designs

Fractional factorial designs are used when there is a large number of possible combinations of variables, and they are generally applied when the factors may occur at two levels. The fraction selected depends on assumed interactions and must come from a statistical design. An example might be a process with six variables occurring at two levels. The total number of interactions is 64, and it is highly unlikely that management would agree to shutting down a line in order to run 64 separate experiments. Information for all two-factor interactions would require only 32 tests. Only 16 tests would be required to investigate some of the two-factor interactions. If no two-factor interaction information were required, only eight tests would be needed. In common practice, an experimenter needs information on only a few of the interactions, and a complete series is rarely run. Fractional factorial plans are useful when the higher-order interactions previously referred to are of no particular importance. This type of plan may be effective when only the main effects of the variables are of interest—that is, when *screening* tests are required.

The eight-test screening plan generates sufficient information to identify which variables (ingredients, for example) have the greatest significance on the outcome of the problem being evaluated. This should further reduce the number of detailed tests required to isolate exact differences. If these eight tests had disclosed that, of the six ingredients in the formula, two were of importance and four had virtually no significance, then subsequent tests need be concerned only with those two ingredients.

Table 12-4 is an example of a fractional factorial designed to measure the response (Y) of variations in eight ingredients at two levels of concentration in a cat food mix. Note the balanced nature of this test plan: The low levels of A (17), which are matched with the low levels of B (7), appear four times; the high levels of A (26), which are matched with the high levels of B (13), also appear four times. Likewise, there are four tests each for the high–low and the low–high combinations of ingredients A and B. Similar two-level pairs are found throughout the test plan for the 8 ingredients and the 16 tests.

From the analysis of variance (Table 12-5), it becomes apparent that ingredients A, D, and E are the ones that truly have a large effect on the response. This screening test has therefore revealed the three ingredients that will require further experimentation.

Table 12-6 consists of several commonly used examples of fractional factorial designs.

Response Surface Designs

After conducting fractional factorial experiments to determine which factors are important, the next step is to explore the interrelationships of these factors using

Table 12-4. Eight-Ingredient Cat Food Test Levels and Responses

Ingredients								Response
A	B	C	D	E	F	G	H	Y
17	7	5	2	0	0	6	3	0.92
26	13	11	5	0	0	6	3	1.44
17	7	5	2	.5	25	12	4	1.43
26	13	11	5	.5	25	12	4	1.91
26	13	5	2	0	25	6	3	1.74
26	13	5	2	.5	0	12	4	1.28
17	7	11	5	0	25	6	3	1.77
17	7	11	5	.5	0	12	4	1.32
26	7	5	5	0	0	12	3	2.03
26	7	5	5	.5	25	6	4	1.46
17	13	11	2	0	0	12	3	1.45
17	13	11	2	.5	25	6	4	1.00
26	7	11	2	0	0	6	4	1.65
26	7	11	2	.5	25	12	3	1.17
17	13	5	5	0	0	6	4	1.67
17	13	5	5	.5	25	12	3	1.29

Table 12-5.   Analysis of Variance

Source	df	MS	F
A	1	0.14251	9.55 $(p < .05)$
B	1	0.00766	0.51
C	1	0.00276	0.18
D	1	0.23281	15.60 $(p < .01)$
E	1	0.74391	49.86 $(p < .01)$
F	1	0.00681	0.46
G	1	0.00051	0.02
H	1	0.00331	0.22
Error	7	0.01492	

response surface designs. The goal here is to find the best combination of the variables.

As an example, consider the pressure, time, and temperature needed to apply a plastic cover to a container to maximize burst strength. In the chapter covering regression analysis, only two variables were involved—the relation between Y and X. In the case of three variables, the model equation becomes more complicated:

Variables: $X_1 = $ Temperature
$\qquad\qquad X_2 = $ Time
$\qquad\qquad X_3 = $ Pressure
Outcome: $Y = $ Burst strength
Model: $\quad Y_{ijk} = B_0 + B_1 X_1 + B_2 X_2 + B_3 X_3 + B_{11} X_1^2 + B_{22} X_2^2 + B_{33} X_3^2 + B_{12} X_1 X_2$
$\qquad\qquad + B_{13} X_1 X_3 + B_{23} X_2 X_3 + E$

where E is the experimental error.

The *central composite design* (see Fig. 12-5) is the classic design for response surface designs. Each side of the cube represents one of the three variables.

The conventional procedure places the lowest level of X_1 at the left, the lowest level of X_2 at the front, and the lowest level of X_3 at the bottom. For example, if X_1 were reduced, the front face would be narrowed vertically.

In previous designs, we have considered only low and high levels of the variables. In the case of response curves, an additional data point is added halfway between the high and low levels, and another set of extra points is added a short distance above the high level and also a short distance below the low level. The extra levels protrude from the faces of the cube and are referred to as *star points.* The low, medium, and high levels are shown with the star points as selected from the data columns in Table 12-7.

If the data columns are examined further (see Table 12-8), it may be noted that the star points are tested against the midpoints of each of the other variables. For

Table 12-6. Fractional Factorial Designs

½ Rep. of a 2^4 Factorial

a	b	c	d	Effects	df
0	0	0	0	Main	4
1	1	0	0	2-factor	3
1	0	1	0	Total	7
1	0	0	1		
0	1	1	0	ab = cd	
0	1	0	1	ac = bd	
0	0	1	1	ad = bc	
1	1	1	1		

¼ Rep. of a 2^5 Factorial

a	b	c	d	e	Effects	df
0	0	0	0	0	Main	5
1	1	0	0	0	2-factor	2
0	0	1	1	0	Total	7
1	0	1	0	1		
0	1	1	0	1	ac = bd	
1	0	0	1	1	ad = bc	
0	1	0	1	1		
1	1	1	1	0	*Note:* Main effects have	
					2-factor aliases.	

½ Rep. of a 2^5 Factorial

a	b	c	d	e	Effects	df
0	0	0	0	0	Main	5
1	1	0	0	0	2-factor	10
1	0	1	1	1	Total	15
0	1	1	1	1		
0	0	1	0	0		
0	1	1	0	0		
0	0	0	1	1		
1	1	0	1	0		
1	0	0	0	1		
0	1	0	0	1		
0	0	1	1	0		
1	1	1	1	0		
1	0	0	1	0		
0	1	0	1	0		
0	0	1	0	1		
1	1	1	0	1		

Table 12-6.   (*continued*)

⅛ Rep. of a 2^6 Factorial

a	b	c	d	e	f	Effects	df
0	0	0	0	0	0	Main	6
1	0	1	0	0	1	2-factor	1
1	0	0	1	1	0	Total	7
0	1	1	0	1	0		
0	1	0	1	0	1	ab = cd = ef	
1	1	1	1	0	0		
1	1	0	0	1	1	*Note:* Main effects have	
0	0	1	1	1	1	2-factor aliases.	

¼ Rep. of a 2^6 Factorial

a	b	c	d	e	f	Effects	df
0	0	0	0	0	0	Main	6
1	1	1	0	1	0	2-factor	7
1	1	0	1	0	1	Error	2
0	0	1	1	1	1	Total	15
1	0	1	1	0	0		
1	0	0	0	1	1	ac = be	
0	1	1	0	0	1	ad = bf	
0	1	0	1	1	0	ae = ac	
1	1	0	0	0	0	af = bd	
0	0	1	0	1	0	cd = ef	
0	0	0	1	0	1	cf = de	
1	1	1	1	1	1	ab = ce = df	
1	0	1	0	0	1		
1	0	0	1	1	0		
0	1	1	1	0	0		
0	1	0	0	1	1		

1/16 Rep. of a 2^7 Factorial

a	b	c	d	e	f	g	Effects	df
0	0	0	0	0	0	0	Main	7
1	1	1	1	0	0	0	Total	7
1	1	0	0	1	1	0		
1	0	1	0	0	1	1		
1	0	0	1	1	0	1		
0	1	1	0	1	0	1		
0	1	0	1	0	1	1		
0	0	1	1	1	1	0		

Table 12-6. (*continued*)

⅛ Rep. of a 2^7 Factorial

a	b	c	d	e	f	g	Effects	df
0	0	0	0	0	0	0	Main	7
0	0	0	0	1	1	1	2-factor	7
1	1	1	1	0	0	0	Error	1
1	1	1	1	1	1	1	Total	15
1	1	0	0	0	0	1		
0	0	1	1	0	0	1	ae = bf = cg	
1	1	0	0	1	1	0	af = be = dg	
0	0	1	1	1	1	0	ag = ce = df	
1	0	1	0	0	1	0	bg = de = cf	
0	1	0	1	0	1	0	ab = cd = ef	
1	0	1	0	1	0	1	ac = bd = eg	
0	1	0	1	1	0	1	ad = bc = fd	
1	0	1	0	1	0	0		
0	1	0	1	1	0	0		
1	0	1	0	0	1	1		
0	1	0	1	0	1	1		

¹⁄₁₆ Rep. of a 2^8 Factorial

a	b	c	d	e	f	g	h	Effects	df
0	0	0	0	0	0	0	0	Main	8
1	1	1	1	0	0	0	0	2-factor	7
0	0	0	0	1	1	1	1	Total	15
1	1	1	1	1	1	1	1		
1	1	0	0	1	1	0	0	ae = bf = ch = dg	
1	1	0	0	0	0	1	1	af = be = cg = dh	
0	0	1	1	1	1	0	0	ag = bh = cf = de	
0	0	1	1	0	0	1	1	ah = bg = ce = df	
1	0	0	1	1	0	1	1	ac = bd = eh = fg	
1	0	0	1	0	1	0	1	ad = bc = eg = fh	
0	1	1	0	1	0	1	0	ab = cd = ef = gh	
0	1	1	0	0	1	0	1		
1	0	1	0	1	0	0	1		
1	0	1	0	0	1	1	0		
0	1	0	1	1	0	0	1		
0	1	0	1	0	1	1	0		

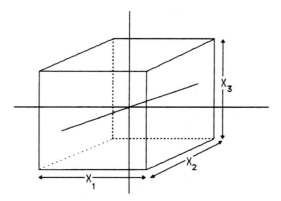

Figure 12-5.   Central composite design.

Table 12-7.   Plastic Lid Response Curve Conditions

	Temperature	Time	Pressure
Star	300	0.5	60
Low	325	0.9	64
Medium	362.5	1.5	70
High	400	2.1	76
Star	425	2.5	80

example, the TIME star points of 0.5 and 2.5 are specified in tests where the TEMP value of 362.5 and the PRESSURE value of 70 are the midpoints of each of these variables. There are instances where the variables cannot be run at the star points because of physical or practical conditions; under these circumstances, the design is modified as needed. A statistician should be consulted when this occurs.

After running the tests and obtaining the outcome information, a linear regression analysis is utilized to determine the equation relating the variables. From the preceding data, a stepwise linear regression analysis produces the following model:

$$Y = 8.925 - 0.00238 * \text{TEMP} - 11.059 * \text{TIME} + 0.0333 * \text{TEMP} * \text{TIME}$$

The remaining B's are nonsignificant. The pressure variable had no effect on the burst strength. Using the equation, it is possible to determine the burst strength for any combination of time and temperature selected. Although only five different levels of variables were examined in the test procedure, the resulting equation permits calculating the burst strength for *any* combination of factors. By constructing a graph of this equation (see Fig. 12-6), the values of all variables can readily be selected.

Table 12-8. Plastic Lid Experiment Design and Outcome

	Experiment conditions		Experiment outcome Y (duplicate tests)	
H_1 Temperature	H_2 Time	H_3 Pressure	Burst strength	
325	0.9	64	8.75	7.5
325	0.9	76	8.25	9.0
325	2.1	64	7.50	7.0
325	2.1	76	8.50	9.5
400	0.9	64	10.50	9.75
400	0.9	76	10.00	10.0
400	2.1	64	12.00	13.75
400	2.1	76	13.00	12.5
300	1.5	70	4.0	7.0
425	1.5	70	12.75	12.0
362.5	0.5	70	8.75	8.5
362.5	2.5	70	11.00	10.25
362.5	1.5	60	9.25	9.0
362.5	1.5	80	9.75	9.5
362.5	1.5	70	9.25	9.5

Figure 12-6. Contour plot of burst strength.
Credit: Wendell Kerr, Calreco, Unpublished manuscript.

Using the chart, select a temperature of about 300°F and a time of 2 seconds; the burst strength will be about 6. A temperature of 350° and a time of 1.5 seconds will produce a burst strength of 9. A temperature of 375° for 0.7 seconds will produce a burst strength of 9. (The curves are read in the same manner as contour maps.)

Whenever a series of experiments is performed to produce a response surface, it is wise to select the best combination of points and retest them to prove that the data are reliable and the calculations correct. In some instances, a response surface analysis will show that more than one set of conditions appears to be optimum. It is again important to verify this information in the real world.

Response surface plans are used where functional relationships are desired and independent variables are quantitative. Variables such as Supplier A and Supplier B (or Machine 37 versus Machine 12) are not suitable for this design. If there are more than five variables in the system, first run a screening design to eliminate some of them, otherwise the testing becomes too large to handle.

Some applications for response surface designs appear relatively trivial but may result in substantial reduction in costs or improvement in quality. In the potato processing industry, peeled potatoes pass over an inspection belt where inspectors remove grossly defective potatoes; trimmers then remove defects after the potatoes have been cut into french fries. If a fixed number of workers is available for defect removal, the question arises: How should the workers be allocated between inspection and trimming? Of 12 workers, should 3 be used as inspectors and 9 as trimmers, or should it be 6 and 6? Using response surface analysis, the optimum allocation may be determined. The variables in such a test would be the number of inspectors and trimmers. The outcome would be the number of defects per pound after both operations.

Mixture Designs

Mixture design is a special response surface type where the total of the variables must equal 100%. In the temperature/time/pressure example, there was no such requirement. Where the variables must equal 100% ($X_1 + X_2 + X_3 = 100$), the general equation

$$Y_{ijk} = B_0 + B_1X_1 + B_2X_2 + B_3X_3 + B_{11}X_1^2 + B_{22}X_2^2 + B_{33}X_3^2 + B_{12}X_1X_2 + B_{13}X_1X_3 + B_{23}X_2X_3 + E$$

becomes greatly simplified to

$$Y = B_1X_1 + B_2X_2 + B_3X_3 + B_{12}X_1X_2 + B_{13}X_1X_3 + B_{23}X_2X_3 + E$$

In Table 12-9, three types of fish are available for use in preparing fish patties: mullet, sheepshead, and croaker. The experimental design is:

Table 12-9. Fish Flavor Experiment Design and Outcome

Experiment conditions			Experiment outcome
X_1 % mullet	X_2 % sheepshead	X_3 % croaker	Y Flavor
100	0	0	4.2
0	100	0	5.1
0	0	100	3.9
50	50	0	3.8, 3.6
50	0	50	4.4, 4.5
0	50	50	4.5, 4.7
33	33	33	4.4, 4.2, 4.3

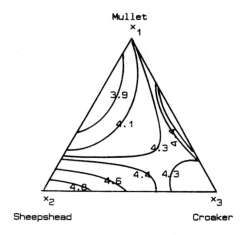

Figure 12-7. Contour plot of flavor.
Credit: John A. Cornell, *Experiments with Mixture Designs, Models, and the Analysis of Mixture Data* (New York: John Wiley & Sons).

Variables: X_1 = Mullet
X_2 = Sheepshead
X_3 = Croaker
Outcome: Y = Mixture flavor
The fitted model is $Y = 4.2X_1 + 5.1X_2 + 3.9X_3 - 3.6X_1X_3 + 1.8X_1X_3 + 0.6X_2X_3$

After running the tests and obtaining the responses, the data are analyzed to arrive at a model equation relating the variables. A graph of the model (see Fig. 12-7) is read as follows: The flat side of the triangle opposite the labeled apex is the zero level of that variable, and the apex is the 100% level for that variable. The center point of the triangle is a mixture containing equal amounts of each of the three fish in the blend. Examining either the model or the chart, it becomes apparent that the better-flavored blends contain higher sheepshead proportions.

One limitation of this type of analysis is that rarely can each ingredient be used at levels from 0% to 100%. It is not possible to go from 0% to 100% sugar in hot cocoa mix, for example. The final cocoa mix must contain cocoa, milk solids, fats, salt, and many other components. In some instances, it is possible to apply limits; in other instances, the total formula might be reduced to some of the key ingredients for purposes of experimentation. In such cases, a final trial run of some of the key findings becomes imperative to test their validity. The need for a final test may be demonstrated in designing tests of dry soup mix, where the salt content may be permitted to vary from 0% to 8%. In the event that the best mixture for soup has a salt content of 6%, actual testing might reveal that this level is too high because of interfering flavor characteristics of other ingredients. It is unlikely, of course, that such a situation would exist near the set points of the actual experiment, because each of these is evaluated for flavor as the *outcome.* At intermediate points, however, there is always a possibility of unusual flavor combinations which should be subsequently confirmed.

Graphical representations of response surfaces become increasingly complex as the number of variables increases; this, among other reasons, accounts for the limitation of five variables for this type of analysis. For example, if a fourth fish were added to the fish patty experiment, the resulting chart would be a three-dimensional tetrahedron, which is dramatically more difficult to interpret.

The following example represents a practical problem which illustrates a central composite design experiment leading to a response surface analysis.

Example
The problem is to set the color specifications on a processed tomato product using the trichromatic color system. The system currently used measures reflectance of the product and expresses the results on a white-to-gray scale which does not always represent satisfactorily the red color perceived by the eye. The trichromatic technique of measuring color provides numerical values for Chroma (brightness), Hue (redness), and Value (lightness to darkness) and is thought to be more reproducible and more characteristic of visual perception. In addition to these objectives, management requested that the specifications be set as simply as possible, preferably by use of a single range of numbers, rather than the three number sets normally used with the newer system. To assist in developing realistic specifications, management has directed that consumer testing be used to establish likes and dislikes of product color.

In the previous examples, the equation terminology was based on B and X relationships. There is no complete agreement on nomenclature. The researcher for this color problem uses a different group of identifiers, but note that the form of the mathematical model is identical.

Consumer response (CR) = f(Hue, Chroma, Value)

$$CR = A_0 + A_1 \times \text{Hue} + A_2 \times \text{Chroma} + A_3 \times \text{Value} + A_{12} \times \text{Hue} \times \text{Chroma} +$$
$$A_{13} \times \text{Hue} \times \text{Value} + A_{23} \times \text{Chroma} \times \text{Value} + A_{11} \times \text{Hue}^2 +$$
$$A_{22} \times \text{Chroma}^2 + A_{33} \times \text{Value}^2$$

In order to develop a continuous scale of consumer response, it is considered necessary to provide each subject with a 9-point hedonic rating scale for evaluation, rather than the two or three points used in previous samples. The rating scale consists of

1—Dislike extremely
2—Dislike very much
3—Dislike moderately
4—Dislike slightly
5—Neither like or dislike
6—Like slightly
7—Like moderately
8—Like very much
9—Like extremely

The central composite design selected for this experiment requires 15 samples to fit the model:

2^3 factorial points	8
Star points above each face	6
Center point	1
Total	15

Figure 12-8 shows the location of the 15 points.

The experimental design (see Table 12-10) is constructed in the same manner as the plastic lid application example. Compare the selection of matched variables in the two examples.

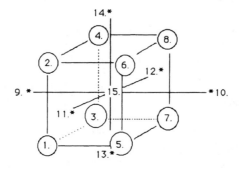

Figure 12-8.   Central composite design.

Table 12-10. Tomato Product Color Experimental Design

Point	Hue	Chroma	Value
1	34	44	20
2	34	44	30
3	34	56	20
4	34	56	30
5	46	44	20
6	46	44	30
7	46	56	20
8	46	56	30
9	31	50	25
10	49	50	25
11	40	41	25
12	40	59	25
13	40	50	17.5
14	40	50	32.5
15	40	50	25

If it were possible, each consumer would be asked to rate all 15 samples and the control; however, fatigue would reduce the sensitivity of the test, and a simpler arrangement is required. By using the balanced incomplete block design, each consumer evaluates eight samples, as shown in Table 12-11. The number of consumers needed to balance the design is governed by the relationship

$$B \times K = R \times T$$

where

B = number of consumers
K = number of samples evaluated by each consumer
R = number of times each sample is evaluated
T = number of samples being tested

For this test, $K = 8$, $T = 15$, and R was specified at 50.

$$B \times 8 = 50 \times 15$$

$$B = 93.75$$

We chose $R = 48$, which yields

$$B \times 8 = 48 \times 15$$

$$B = 90$$

Table 12-11.   Consumer Responses

Consumer no.	I	II	III	IV	V	VI	VII	VIII
				Sample presentation order				
1	11	4	2	5	10	3	14	15
2	4	1	3	15	13	11	6	9
3	9	2	14	4	7	1	10	13
4	15	3	1	10	8	13	7	5
5	7	13	10	12	11	2	3	6
6	6	10	12	1	4	14	8	3
7	12	9	13	14	15	10	11	8
8	10	11	9	7	6	8	5	4
9	5	14	7	3	1	9	12	11
10	8	15	6	2	3	7	9	14
11	1	7	11	8	2	4	15	12
12	2	6	15	9	5	12	1	10
13	13	8	5	11	14	6	2	1
14	14	5	4	6	12	15	13	7
15	3	12	8	13	9	5	4	2

Note: T = 15 samples; k = 8 samples/consumer; B = 15 consumers.

In order to meet the required number of consumers (B = 93.750), it will be necessary to repeat the design six times ($15 \times 6 = 90$), randomizing the order of sample presentation. The test is completed and the average response for each sample calculated and tabulated. The consumer hedonic rating for each sample is tabulated in part in Table 12-12.

To simplify the statistical analysis, it is necessary to eliminate one column of variables. The variables removed from the analysis for the present are the data for values. This, in effect, reduces the cube model to a square, with 9 responses to consider instead of the 15 available. These 9 responses (plus the 5 center-point replicates) are selected and shown in Table 12-13 as a subset. Following the columns of data, the consumer response mathematical model is shown with the value variables omitted.

The simplified model is then solved using data from Table 12-13.

$$\text{Consumer response} = A_0 + A_1 \times \text{Hue} + A_2 \times \text{Chroma} + A_{11} \times \text{Hue}^2 + A_{12} \times \text{Hue} \times \text{Chroma} + A_{22} \times \text{Chroma}^2$$

Using a regression analysis computer program, the following equation is obtained:

$$\text{Consumer response (CR)} = -52.23 + 1.16 \times \text{Hue} + 1.20 \times \text{Chroma} - 0.01336 \times \text{Hue}^2 - 0.01090 \times \text{Chroma}^2$$

Table 12-12. Average Consumer Responses

Sample no.	Hue	Chroma	Value	Average response
1	34	44	20	4.18
2	34	44	30	3.19
3	34	56	20	5.49
4	34	56	30	5.08
5	46	44	20	4.38
6	46	44	30	4.69
7	46	56	20	5.43
.	.	.	.	.
.	.	.	.	.
.	.	.	.	.

Table 12-13. Simplified Model with Value Responses Omitted

Sample no.	Hue	Chroma	Response
1	34	34	4.18
2	34	56	5.49
3	46	44	4.38
4	46	56	5.43
5	31	50	3.23
6	49	50	6.21
7	40	41	3.74
8	40	59	6.10
9	40	50	5.62
10	40	50	6.10
11	40	50	5.66
12	40	50	5.61
13	40	50	5.76
14	40	50	5.91
15	40	50	5.91

or

$$0.01336(\text{Hue} - 43.4)^2 + 0.0109(\text{Chroma} - 55)^2 = (5.9 - \text{CR})$$

Since CR = 5 = neither like nor dislike, the right-hand term becomes approximately $6 - 5 = 1$. Therefore,

$$\frac{(\text{Hue} - 43.4)^2}{67.37} + \frac{(\text{Chroma} - 55)^2}{82.57} = 1$$

The best responses are Hue at 43.4 and Chroma at 55.0.

As we have seen before, one of the advantages of this type of analysis is that the intermediate points of consumer preference may be extrapolated without conducting another series of experiments. The easiest way to do this is graphically. The graph of the response surface is approximated in Figure 12-9. The intersection of the two axes has a value of 43.4 Hue and 55.0 Chroma, with a hedonic response of 6.5 (the maximum preference rating of all of the consumers). The inner ring would have a value of 6.0, the intermediate ring 5.5, and the outer ring 5.0. It is therefore possible to interpret consumer preferences at any combination of values.

Figure 12-9 shows a small portion of the total response surface and may be thought of as the upper surface with the observer standing above it and looking down. A computer printout of the entire surface is shown in Figure 12-10 and represents a three-dimensional picture.

Returning to the initial statement of the experimental design goal, a single range of numbers may be selected as a color specification. An initial candidate for such a specification might be: If a consumer rating is above 5.5, accept the sample; if the score is below 5.5, resample. If resample is above 5.5, accept; otherwise, reject.

The question of what to do if the value variable were also significant should be mentioned. If value were included along with Hue, Chroma, and Response, a mathematical solution would require a four-dimensional model—extremely difficult to depict! It would be wise to consult with a statistician, who might propose a compromise. One compromise method is to use only two of the variables and the response (as we have done here), then extract a slice from the model which includes the required specification levels and repeat the experiment (in whole or in part) with value, response, and one of the other variables. For a complete test, a second replicate test would be required with the remaining variable. If the color problem is critical, the entire series of tests will be necessary; otherwise, a few spot values may be tested to rough out the location of specification levels for all the variables.

Elliptical contours are particularly useful because they define maximum responses. There are many other contours that are less desirable. One of these is the so-called saddle point response (see Fig. 12-11). The intersection of the axes is

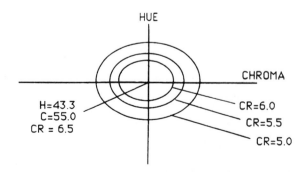

Figure 12-9.   Surface response, elliptical contours.

Figure 12-10.   Complete response surface.
Credit: Randall Hamllin, *The Place of Design in Process Improvement* (Hunt Wesson Foods, Inc., 1987), K-116.

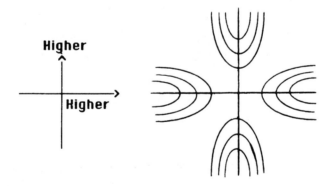

Figure 12-11.   Saddle point contours.

not the maximum response, but rather an intermediate value from which moving in a vertical or right horizontal direction shows increasing responses; conversely, moving down or to the left shows decreasing responses.

EXPERIMENTAL DESIGN ANALYSIS BY CONTROL CHART

The classical techniques of analyzing the results of the tests are based on analysis of variance. It is also possible to use a different approach in which quality control techniques, such as control chart analysis, analyze the outcomes. To illustrate, let us look at a microwave bread test similar to the one examined earlier. Four formulas will be tested in four different microwave ovens (all of which are 600+ watt rated). As before, the outcome of interest is the volume of the loaves.

The test procedure calls for baking 5 loaves of each formulation in each oven, for a total of 20 loaves per oven. (As a matter of interest, the number of samples selected was five because the distribution of the means of five samples is very close to normal, regardless of how skewed the distribution of individual samples might be. This is less true for smaller-sized samples, and larger sample sizes do not appreciably improve the normality of the averages.) After baking, the average volume of the loaves is measured, and the averages of each group of five samples is calculated. In addition, the standard deviation of each group of five is calculated. To summarize, the means and standard deviations for each group of five are to be calculated for each microwave, and all 16 sets of data analyzed. Table 12-14 represents the outcomes of the baking tests for four different formulations in four different microwaves. A final series of calculations yields the overall average for each formulation and for each oven.

Merely looking at the data, it appears obvious that Formulation 4 has the highest volumes in each oven. What is less obvious is the reliability of this conclusion. Can one be sure that if the tests were repeated several times, the volume of Formulation 4 would always be greater than the others? To rephrase the question: Can one state with any degree of confidence that the results for Formulation 4 are significantly different, reproducible, and higher than the others? Some kind of an analysis must be conducted at this point. t-Tests are useful for comparing differences between two means or comparing a mean to a standard. When there are three or more to compare, the analysis of variance technique may be used. This uses the F-ratio (analysis of means):

$$F\text{- ratio} = \frac{\text{Variance between means}}{\text{Variance between each sample}}$$

Table 12-14. Effect of Microwave Oven on Bread Formula Volume

Formulation	Microwave oven				Overall average
	GE	TA	SA	AM	
1	156/19.6	143/14.7	160/21.3	137/25.4	149.0
2	150/11.9	151/12.3	157/15.9	129/9.1	146.8
3	112/30.1	137/24.6	126/25.9	109/19.0	121.0
4	162/19.1	151/18.7	175/15.4	141/20.1	157.2
Oven average	145.0	145.5	154.5	129.0	143.5

Note: Average volume of five loaves, measured in cubic inches.

If the variance between the means is greater than the variance within the means (between samples), then there are significant differences. This requires many statistical calculations. Note that, for oven TA, the volume means for Formulations 2 and 4 are equal at 151, and the variation for Formulation 2 (12.3) is much less than that for Formulation 4 (18.7). In other words, if only one oven, TA, had been used for this test, Formulation 4 would not have been so obviously best, as was stated in the preceding paragraph. This emphasizes again the need for a balanced test.

Eliminating the differences between microwaves might be accomplished by using block designs. This method has already been examined. But there is another, perhaps simpler, method which uses a control chart. Obviously, we are not dealing with a continuous process which lends itself to control chart techniques, but we *are* dealing with a number of values, the relationship of which can be demonstrated graphically. The uniformity of the data can be represented for visual, rather than mathematical, evaluation. This technique was suggested years ago (before the common availability of computers) and presented by Ellis Ott in his book *Interpretation of Data* (1975). This technique is still an excellent way to examine data.

A control chart is constructed using the data from Table 12-14. The center line is the overall average loaf volume—143.5 cubic inches. Limits are calculated which are the equivalent of three standard deviations, such as those used on conventional control charts. The calculation of the limits will be explained shortly. First, let us interpret the chart data (see Fig. 12-12).

Examine the data for uniformity (consistency) of loaf volumes for each oven. Formulation 1 is fairly uniform; all loaf volumes are within the control limits and equally distributed on both sides of the average. Formulation 2 appears satisfactory. However, two of the four loaves baked with Formulation 3 are below the lower control limit, indicating that this may be an unsatisfactory formula. Formulation 4 shows three readings within the control limits, and one loaf volume higher than expected from chance alone. Based on these observations, Formulation 4 appears to be the candidate to be taken to the next stage of development.

Figure 12-12.   Control chart analysis of four ovens.

Could one expect to find similar results if this test were repeated? A statistical test is required to determine whether the selected formula is significantly different from the others. This is accomplished by performing the analysis of means. The calculations are as follows:

$$\text{Average } \overline{X} = \frac{(\Sigma X)}{N} = 143.5 \text{ cubic inches}$$

$$\text{Average variance } S = \sqrt{\frac{S_1^2 + S_2^2 + S_3^2 + \cdots S_{16}^2}{16}} = 19.7$$

Each cell represents ($n = 5$) loaves of bread; therefore, there are $n - 1$ or 4 degrees of freedom in each cell. There are 16 sets of 4, or 64 degrees of freedom.

Control charts use three standard deviations as the control limits. Determining three standard deviations for process control would normally require the use of the Z table, which assumes an infinite number of observations. Here we shall use a slight modification of that principle because, in this example, several means (16) are involved, but the data are limited—it does not represent an ongoing process. The control limits to be used are expressed as *decision lines* and are found using the following formula:

$$\text{Limits} = X \pm H_\alpha \hat{\sigma}_{\overline{X}} \quad \text{where} \quad \hat{\sigma}_{\overline{X}} = \frac{\sigma}{\sqrt{n}}$$

$$= 143.5 \pm 2.94 \left(\frac{19.7}{\sqrt{5}} \right)$$

$$= 143.5 \pm 25.9 \text{ or } 117.6 - 169.4$$

The value for H_α is found in Table A-12 in the Appendix by extrapolating for $k = 16$ and df = 64. The table includes values for H_α at a producer's risk of rejecting a good lot (α) of 5%; that is, there is a 5% risk of finding a significant difference between means when there actually is none.

As mentioned previously, the usual method of analyzing data when a computer is available is to have a technician enter it into an analysis of variance program and let the computer do the work. By using the graphical method described here, however, one retains the feel for the data. The pictured data shown on a graph should be much more meaningful to an experimenter than a numerical computer printout.

More information may be extracted from the data. It is possible to determine if the differences observed between formulas and the differences observed between ovens are statistically significant. The process of calculating the significant difference level is called the *multiple range test for the just significant difference value* (JSV) and uses the formula

$$JSV = qx\sqrt{\frac{s^2}{n}}$$

where q is found in Table A-11 in the Appendix: the upper 5% of the points of the studentized range $q = (x_n - x_1)/s$, which is in the Appendix.

$$JSV = 3.74\sqrt{\frac{19.7^2}{5}} = 32.9$$

The differences between ovens must be at least 32.9 cubic inches to be significant. The largest differences between ovens as shown in Table 12-14 are

For Formulation 1: SA – AM = 160 – 137 = 23
For Formulation 2: SA – AM = 157 – 129 = 28
For Formulation 3: TA – AM = 137 – 109 = 28
For Formulation 4: SA – AM = 175 – 141 = 34

It may be concluded, therefore, that there is a significant difference between oven SA and oven AM, but no others. In other words, Formulation 4 is more adversely affected by choice of ovens than are the other formulations.

Another important comparison is the significance of differences between formulations:

For oven GE: 4 – 3 = 162 – 112 = 50
For oven TA: 2 – 3 = 151 – 137 = 14
For oven SA: 4 – 3 = 175 – 126 = 49
For oven AM: 4 – 3 = 141 – 109 = 32

The conclusion here is that Formulation 3 has a significantly lower volume than Formulation 4 for three of the four ovens. It is not clear whether Formulation 4 is better than the other two. The protocept to select for subsequent testing would probably be number 4, but it is indicated that Formulations 1, 2, and 4 should be tested once more in a single oven (to rule out the effect of ovens on the outcome). From a marketing point of view, oven SA would be selected, because it appears to contribute to the highest volume for Formulation 4.

The preceding discussion shows the advantage of using this multiple range test rather than the analysis of variance (ANOVA) to evaluate the formulations and ovens. Both types of analysis permit the use of a number of variables in a single test, thus showing the interactions of variables and saving considerable time and money compared to more traditional single-variable testing procedures. ANOVA tells the experimenter whether the variation between means is significant or not, but it fails to show which of the means are the significantly different ones. The multiple range test accomplishes both of these objectives. An additional advan-

tage of the multiple range test is that the final report to management is both pictorial and tabular—easier to present and understand.

The following list summarizes the information required to design an experiment:

1. Definition of the purpose and scope.
2. Identification and specification of the experimental variables.
   (a) Primary variable
   (b) Background variables (control by blocking)
   (c) Uncontrolled variables (control by randomizing)
   (d) Variables held constant.
3. Estimates of repeatability (estimate of error or standard deviation).
4. Desirability of conducting experiment in stages.
5. Prior knowledge about results.
6. Constraints on the variables.

We have explored a few of the more commonly used experimental designs, but
have merely skimmed the surface of the accompanying mathematical manipula-
tions of data. Although computer programs are available to perform this portion
of the designs, it is suggested that the manual procedures should be mastered in
order to understand which programs are applicable to the problem at hand.

13 Vendor Quality Assurance

It is difficult to generalize about raw material costs in the food industry because the products range from flavored water to caviar. The cost of an inexpensive drink's flavoring, sweetener, acidifier, colorant, and packaging might be as little as 15% of the factory door cost. Restaurant operations may estimate that 40% of the total operation cost is for purchased foods; high-ticket food items could exceed 50%. A high-cost raw material focuses attention on the need for uniform acceptable quality of that material to a greater extent than a modestly priced unprocessed food ingredient does. From a quality control standpoint, this is an unfortunate concept because the final quality of the finished product depends not on the price of the raw ingredients, but on their quality level and uniformity. With this in mind, it seems that a large part of the responsibility for the successful production of a uniform quality food item, manufactured with a minimum of scrap and rework, rests with the purchasing department. In fact, the responsibility rests with all of the processor's departments—each of which should be in close contact with suppliers.

There is an attempt by some to refer to the quality at all steps in the construction of a product as the responsibility of each worker who performs a function affecting the finished quality, and that each worker, therefore, may be considered to be a vendor or supplier to the worker at the next step in the process. This philosophy suggests that all workers along the chain—from raw material to the final product—are also customers of the vendors in the preceding operation. "Quality is everyone's business" may be a perfectly valid slogan, but for the sake of clarity, we shall consider vendors (or suppliers) the organizations who supply raw materials to the processors (or vendees, or customers, or manufacturers). Vendors may be farmers or manufacturers of cans, corrugated board, process machinery, kraft bags, food additives, or any other material or service that is involved in the flow of a product through the processor's plant.

Figure 13-1 shows a typical flow of raw materials to the processor and may help identify some costly flaws of commonly used systems.

Ordinarily, the manufacturing department issues a purchase requisition for raw

Figure 13-1. Movement of purchased goods from vendor to processor.

materials or supplies, usually with a description or specification, and this is where quality control problems might originate. Unless the material quality requirements are clearly defined when forwarded to the purchasing department, unusable material could be ordered. The purchase order flows to the vendor's customer service department which reviews (or, at the very least, glances at) the requirements and prepares either a ship-from-stock order or a production schedule for the vendor manufacturing facility. In the course of production, the vendor's quality department identifies product that does not conform to the requirements and directs it to rework or scrap. The rest is shipped to the processor with a bill of lading and possibly a certificate of quality. When received by the processor, the material is usually inspected for compliance with the specification before releasing it for production. Inspection may range from merely counting the number of cases received and comparing it with the quantity ordered, to sampling inspection or even 100% inspection and analysis before approval. Again, unsatisfactory material is either reworked or returned to the vendor for credit.

This simple path has been used for years by food processors of all sizes, many of whom are well aware of its inadequacies. Accepting raw materials of less than satisfactory quality practically guarantees that the finished product will be of less than satisfactory quality as well, or that it will have satisfactory quality at higher production costs due to sorting, scrap, and rework. Incomplete specifications result in frustration and expense, even with duplicate testing at the supplier's and processor's plants. Consider a purchase order calling for another shipment of 50,000 cans with the usual red company logo. Perhaps the vendor's quality control lab notices that the color seems a little lighter than they remembered it from the last run but assumes it is close enough. When the cans are received, the processor's quality control lab notices that the color is a little less red than the last run, but, as there are no more cans available to run, the production department decides to use them rather than shut down the line. When the off-color cans are shipped from the plant to the field warehouses, to the distribution centers, and finally to the stores, the processor's salesmen see them for the first time, and angrily demand that these substandard orange cans be replaced with correct red-colored ones. When the cans are stacked on grocers' shelves next to normal-colored cans, the resulting contrast clearly indicates that the company is unconcerned with controlling quality, and sales are bound to suffer.

This hypothetical situation might have been halted at the outset with a color requirement specified in numerical terms (and including acceptable limits) as part of the purchase order; and it could have been stopped at either the vendor's quality control laboratory, or at the processor's quality control inspection at the receiving dock, or as the cans were first fed to the production lines. A number of people had the opportunity to take effective action after observing the defect. It also might have been stopped if an open line of communication existed between the purchasing and customer service departments or between vendor and processor quality control departments. Perhaps the production department should not have the authority unilaterally to decide whether it is more economical to run substandard materials or to shut down a production facility.

There is another lesson to be learned here as well. If the vendor's quality control laboratory is required to examine the cans, why is it necessary to duplicate this effort at the processor's quality control laboratory? The answer for this example is simple: because the two laboratories do not talk to each other. A spot check and occasional audit by the processor's laboratory might be all that is needed, provided that the vendor knows exactly what is required and has an open line of communication to the processor's laboratory in the event of nonstandard product. If such a system could be implemented, the size of the quality control inspection crew presently assigned to incoming inspection could be reduced. Simply by improving communications, we have already improved productivity!

VENDOR–VENDEE RELATIONS

Obviously, testing and retesting do not necessarily prevent quality problems, nor do they cement relations between vendor and customer. There are opportunities to improve both of these situations. When properly designed, a program known as *vendor certification* can be an effective tool. This relationship cannot be proclaimed and put into use overnight; it must be developed slowly. Philip Crosby has invented the "Quality Management Maturity Grid" (Crosby, 1979), which lends itself to this growth of trust and understanding between suppliers and customers.

> *Stage 1: Uncertainty.* Orders are placed with the vendor and accepted without question by the purchaser. Little but the most obvious receiving inspection is performed.
>
> *Stage 2: Awakening.* The purchaser becomes aware that much material is rejected on the line due to drifting quality of the raw material. He suspects that there must be a better way of getting uniform and acceptable quality.

Stage 3: Enlightenment. Sampling inspections are increased and quality engineering studies are conducted to explore improvement of raw material quality uniformity. Exploratory discussions start between vendor and supplier.

Stage 4: Wisdom. The purchaser formalizes incoming inspection plans. Some progress is made toward vendor certification of analyses of shipped lots. The purchaser is aware of the possibility of selection and certification by audit of candidate supplier's plants to ensure that the supplier's plant is in statistical control and is capable of producing acceptable product.

Stage 5: Certainty. Acceptance of shipments is conducted solely by the terms of the purchase order.

According to Kaoru Ishikawa, there are 10 quality control principles for good vendor/vendee relationships.

1. Both vendor and vendee are fully responsible for the quality control applications with mutual understanding and cooperation between their quality control systems.

2. Both vendor and vendee should be independent of each other and respect the independence of the other party.

3. The vendee is responsible for bringing clear and adequate requirements to the vendor so that the vendor can know precisely what he should manufacture.

4. Both vendor and vendee, before entering into business transactions, should require a rational contract between them with respect to quality, quantity, price, delivery, terms, and method of payment.

5. The vendor is responsible for the assurance of quality that will give satisfaction to the vendee, and he is also responsible for submitting necessary and actual data upon the vendee's request.

6. Both vendor and vendee should decide beforehand the evaluation method of various items which will be admitted as satisfactory to both parties.

7. Both vendee and vendor should establish, in their contract, the systems and procedures through which they can reach amicable settlement of disputes whenever problems occur.

8. Both vendee and vendor, taking into consideration the other party's standing, should exchange information necessary to carry out better process/quality control.

9. Business control activities (production, inventory planning, ordering, reporting systems) should be performed sufficiently to maintain their relationship on an amicable and satisfactory basis.

10. Both vendor and vendee, when dealing with business transactions, should always take into full account the consumers' interests.

In the words of Kaoru Ishikawa, "Always have a mutual confidence, cooperation, and a degree of live-and-let-live in vendor/vendee relations" (Ishikawa, 1985).

SPECIFICATIONS FOR RAW MATERIALS, INGREDIENTS, AND SUPPLIES

By now it should be clear that vendor quality starts with specifications. Many food-related specifications exist, occasionally accompanied by detailed test methods, in government and commercial publications.

FEDERAL SOURCES
Food and Drug Administration
National Institute of Standards and Technology (formerly Bureau of Standards)
U.S. Department of Agriculture
Military Specifications
General Services Administration

INTERNATIONAL ORGANIZATIONS
Food and Agriculture Organization (FAO)
World Health Organization (WHO)
Codex Alimentarius
International Organization for Standardization (ISO)

INDUSTRY SPECIFICATIONS (STANDARDS AGREED UPON
BY PRODUCERS IN AN INDUSTRY)
American Dairy Products Institute
American Egg Board
American Frozen Food Institute
American Institute of Baking
American Meat Institute
American Spice Trade Association
Associations for Dressings and Sauces
Can Manufacturers Institute
Composite Can and Tube Institute
Flexible Packaging Association
Glass Packaging Institute
Grocery Manufacturers of America
Institute of Food Technologists
International Coffee Organization
National Coffee Association
National Dairy Council
National Soft Drink Association
Processed Apples Institute
Technical Association of the Pulp and Paper Industry
The Food Processors Institute
The Vinegar Institute

SPECIAL PURCHASING SPECIFICATIONS
Standards agreed upon between vendor and vendee

Some care should be taken when selecting standardized specifications for the first time. They may be more stringent than needed, or they may not be sufficiently definitive to satisfy the requirements of a particular process. A moisture specification worded "not to exceed . . . %" may be required to ensure satisfactory keeping quality, but perhaps a lower limit might be required to permit rapid rehydration or prevent clumping when the material is used in a mix. With the exception of government standards (where the entire specification must be adhered to), a specification for raw material should be as simple as possible, including only those attributes considered critical to the user's process. As a general rule, the fewer critical specifications demanded by the purchaser, the lower the price from the vendor. There is little point in specifying the whiteness of corn sugar used in a mix destined for black licorice candy ropes.

A number of factors affect the longevity of raw material specifications. Some become completely obsolete as substitutes or new varieties become available. Public attitude and demands change, customers change their preferences, new technology makes it possible to produce new foods or modify existing ones, and modified raw materials may become available as established ones become scarce.

The "perfect" specification can be created without any familiarity with the product, but, to be of any value, a specification must be practical and workable. It is very difficult to write a specification that can be used on the factory floor to produce goods or to purchase raw materials on a widely available basis. Whenever possible, specifications should be developed mutually between the vendor and the vendee—so that the test methods, shelf life, sampling methods, and storage conditions are compatible to both parties' operations.

Many requirements for preparing specifications were covered in Chapter 7. For the sake of completeness, a checklist of frequently used specification parameters for raw materials follows:

- Bacteriological Standards
  Standard plate count, yeast, mold, coliforms, *E coli, Salmonella, Lactobacillus,* thermophilic sporeformers, and so on (usually expressed in maximum numbers per gram, or presence or absence).
- Physical Standards
  Particle size, flowability, visual color, solubility, bulk density, flavor, odor, sediment, and so on.
- Analytical/Chemical Standards
  Fat, moisture, water activity, volatile oil, starch gelatinization, pH, sugar content, residual pesticides, sodium, protein, and so on.
- Sanitation Standards
  Insect fragments, rodent hairs, foreign material, special processing conditions, personnel hygiene, procedures for plant and equipment sanitation, hygiene, pest control, and so on.
- Other Standards
  Shelf life, storage conditions, product coding (lot numbers, item numbers, dates), handling procedures, formulas, and so on.

- Sampling Procedures and Test Methods
  Type of sampling (random, stratified, periodic, etc.), number of samples
  per lot, per period of time, per shipment, specific test methods (AOAC,
  FDA, nonstandard, etc.), lot acceptance sampling and evaluation.

Specifications for packaging materials require dimensional data, as well as
packaging composition (board, laminates, film, adhesives, cans, jars), printing
detail (graphics, logo), and sometimes chemical and physical standards (oxygen
permeability, moisture barrier, product transfer resistance, or other chemical barriers).

QUALITY ASSURANCE OF PURCHASED GOODS

When one considers the astronomical number of materials, suppliers, processors,
products, and materials in the food processing industry, it is no wonder that no
single set of quality control procedures has been evolved to satisfy the vendor, the
processor, and the final customer. Plans have been made available on a regular
basis for many years, starting as early as 1948 with H. F. Dodge's discussions on
avoiding lot rejections triggered by sampling plans (Dodge, 1948). Even the basis
for inspection—whether it should be by variables or by attributes—has been the
subject of heated debate since the early 1940s. Some maintain that attribute inspection is easier, requiring relatively unskilled personnel, speedy examinations,
and ironclad accept-or-reject criteria. Others argue that there are at least three
classes of defects for each attribute and that considerable judgment is required to
distinguish among them. Unsolved is the problem of when it is more economical
to return a substandard shipment, to subject it to 100% screening inspection, or to
use it and charge back to the vendor the added costs of using the substandard material on the production line. On the other hand, even though variables analysis of
a shipment may require skilled personnel, much time, and many more examinations than an attribute-based system does, the results of variables testing are concrete, indisputable numbers requiring no judgment on the part of the technician,
and they rarely cause friction between the vendor and the vendee.

One would expect that a logical analysis balancing the costs and the savings of
each alternative would produce a series of plans at various end-product quality
levels which would satisfy the vendor, the vendee, and the final consumer. In the
real world, this is merely the starting point. The vendor wishes to maximize profits, the processor wishes to minimize costs, and the consumer looks for perfection
in the product. It is not possible to negotiate, and usually risky to compromise, the
level of product quality offered to the customer. Negotiations between vendor and
vendee are the only realistic tools available. There has to be a starting point from
which continued improvements can be made.

There are hundreds of acceptance plans available; the choice should be based
on mutual agreement between the seller and the buyer. One attribute plan in use
for years is based on Military Standard 105E of the U.S. Department of Defense.

It offers three levels of inspection for acceptance—normal, tightened, and reduced—and provides detailed switching procedures between the levels based on lot rejection experience. It requires that lot inspection be initially set at *normal,* changed to *tightened* when 2 out of 5 consecutive lots or batches have been rejected, returned to *normal* when 5 consecutive lots are again found acceptable, lowered to *reduced* when 10 consecutive lots are found acceptable, and so on. As cumbersome as this appears, it has worked successfully for years in the quality control acceptance function of military and civilian industries alike. It has found its way into process control as well as material control and has been used successfully as a reliable (although not particularly rapid) indicator of quality level changes.

One of the goals of acceptance sampling is to entirely avoid raw material testing at the processors' plant. The thought behind this goal is that the vendor is far more experienced in testing the line of products he sells than the processor is, and he may be in a better position to correct defects as they are uncovered. If this generates added costs for the vendor, they may be passed along to the processor, frequently at less cost than adding a second level of quality inspection. The processor would then be in a position to receive certified shipments, along with quality control documents showing test data, control charts, and final reports pertaining to each shipment. He would reduce his raw material quality inspection system to an occasional audit, preferably at the vendor's operation. A word of caution: Where critical defects might be found in a raw material, certification is desirable, but it is not sufficient evidence of compliance. Audit samples must be run on each shipment, regardless of efforts on the part of the supplier. The responsibility for a safe food product rests with the processor, not the raw material vendor.

An interesting plan suggested by Ishikawa shows the interaction between vendor and vendee and shows the developing and maturing relationship. This plan is outlined in Table 13-1.

In Phase 1, quality control is least developed. The vendor ships the raw material as soon as it is manufactured or assembled, without any inspection or testing. The vendee accepts the material without inspection and sends it to the manufacturing operation. Manufacturing then has to 100% inspect the material to select that portion that is suitable for production. In some cases, manufacturing may not even inspect the material but may use it blindly. Because this generates unsatisfactory finished product, low productivity, and excessive scrap and rework, the vendee proceeds to Phase 2, 100% inspection at the receiving dock. If this inspection procedure is effective, only acceptable materials now reach the manufacturing operation, but the added inspection costs may equal or even exceed the cost of lower productivity in Phase 1.

Prompted by the volume of returned goods from the vendee's receiving inspection, the vendor now engages in 100% inspection of finished goods, and the vendee's quality control notes an improvement in the quality of the materials received. As a result, the vendee reduces receiving inspection from 100% to a statistical sampling plan, such as Military Standard 105E (Phase 4). If he is more confident and if the vendor is willing, the vendee will now be satisfied to merely

Table 13-1. Development of Vendor–Vendee Relationship

	Vendor		Vendee	
Phase	Manufacturing dept.	Quality dept.	Quality dept.	Manufacturing dept.
1	—	—	—	100% inspection
2	—	—	100% inspection	—
3	—	100% inspection	100% inspection	—
4	—	100% inspection	Sampling or checking inspection	—
5	100% inspection	Sampling inspection	Sampling or checking inspection	—
6	Process control	Sampling inspection	Checking or no inspection	—
7	Process control	Checking inspection	Checking or no inspection	—
8	Process control	No inspection	No inspection	—

review the vendor's quality control records. Otherwise, the vendee might try reduced inspection or skip-lot sampling.

The vendor is now encouraged by the reduced costs brought about by a minimal quality control effort and he expands the coverage (Phase 5) to include 100% inspection in his own processing department while reducing the finished material sampling to an acceptance sampling plan prior to shipment. The vendee, meanwhile, continues minimal incoming inspection.

Although he may have reached a rewarding level of shipping all satisfactory material, the vendor should explore the possibilities of reducing his own scrap and rework by using some of the tools of experimental design and quality engineering. By studying and improving his process capability, the vendor should be able to replace 100% process inspecting with process control techniques while still retaining the finished goods sampling method for lot acceptance. The vendor has reached Phase 6. The vendee continues to check incoming material quality but may have reached the point where shipments, accompanied by a certificate of analysis, can be accepted.

In Phase 7, the effectiveness of process control at the vendor's operation allows reduction of end-of-the-line sampling inspection and lot acceptance to checking a few samples. The vendee may now accept materials which are accompanied by the vendor's process control charts, letters of analysis, or similar certificates.

Phase 7 is probably attainable; Phase 8 is more a goal than an actual condition. Phase 8 assumes an ideal situation where process capabilities have been improved and reliable day-to-day process control has been implemented. Under these conditions, shipping inspection by the vendor is no longer required. It is as if the vendor's manufacturing line had become attached to the start of the vendee's processing line. It is unlikely that this condition can be maintained—if reached at all—because employees and management change, people become comfortable at their work and make mistakes, and new and improved raw materials are always demanded by processors. By keeping in constant contact with the vendor, however, these conditions can be predicted before they develop, and corrective steps taken.

It should become apparent that, as both vendor and vendee strive to approach Phase 8, both parties may drastically reduce the number of inspection personnel required. This reduction, as well as the cost reduction from the elimination of defects, is accompanied by a rise in productivity and the establishment of reliable vendor quality assurance.

Ishikawa's philosophies are somewhat detached from people. The statements appear as absolutes and as goals to be attained by vendors, vendees, companies, departments, and so on. This may be a misinterpretation by Ishikawa: Within many companies, there exists an "us-and-them" attitude, which is extremely counterproductive. Quality control and production departments frequently have conflicting goals (*produce according to specifications or shut down* versus *produce at least cost and get it out the door on time*). We have already looked at the conflicting goals of the vendor and his customer. One might expect the background, training, interests, and goals of purchasing personnel to be far less technical than those of quality control professionals. Somehow, all of the suspicions, antagonisms, conflicts, and jealousies of these parties have to be put aside in order for a quality control system to work. Imagine the devastating impact of a statement such as, "You, Mr. Vendor Quality Manager, really made a big blunder on that last shipment to us, but we (Vendee Quality people) discovered it in time." One can imagine Mr. Vendor Quality Manager thinking up some form of retribution to darken the reputation of his counterpart at the vendee's plant.

To improve productivity by purchasing defect-free raw materials, the relationship between quality and purchasing is vital. If the purchasing department has not been exposed to the details and the power of X-bar and R charts, they are unlikely to look upon such charts favorably when offered by quality control. If the quality department is unaware of the difficulty of negotiating lowest prices with independent vendors while still bowing to company management's intuitions, prejudices, and favorite suppliers, they may not be anxious to work with the quality department of a less-than-desirable vendor selected by the purchasing department. Such situations are common. Although not completely avoidable, they can be smoothed out if both parties earnestly work toward the common goal of productivity.

SELECTING AND NURTURING A SUPPLIER

The responsibility of vendor selection has historically rested with the purchasing department. Somewhat archaic criteria are still used in the process: service, reputation, and lowest price. Over the years, selections have been affected to varying degrees by promotional efforts on the part of potential suppliers. Although it seems ridiculous, decisions have been swayed by friendships, golf, gratuities, and dinners. Such suspect practices continue because of the philosophy, "If it ain't broke, don't fix it." If quality and productivity are adversely affected by such selections, then the supplier system *is* broke and should be carefully reassessed. Fortunately for both the food industry and the consumer, these practices are being replaced by far more logical decision-making criteria, with the selection process including input from other concerned department managers.

If a price from a potential vendor appears completely out of line with the competition, the reasons for the higher price should be investigated before outright rejection. There is always the possibility the price could include special services, higher or more uniform quality, warehousing, delivery concessions, and so on. Price used to be the first topic of discussion, but now it has moved further down the list. The first question should be whether a potential vendor is capable of supplying precisely those raw materials, ingredients, or packaging supplies required by the vendee's specifications on a rigorous delivery schedule. Close cooperation between the two parties may be required in the event that a vendor is unable to supply material to the specifications of the processor. In such a case, it is desirable to have selected a vendor who will develop willingly a satisfactory material with the assistance of the vendee. An easily accessible line of communication should be available. The vendor's use of process control and the ability to respond to the need for enhancing process capabilities should be explored. In addition, the vendor should be able to control production volume and have the financial ability to ensure expansion if required to meet increasing vendee needs.

A selected vendor should have a stable management system that is respected by others in his industry. The vendor should understand the vendee's management philosophy as well, and close cooperation between the two managements should be established and maintained. Other important considerations in choosing a vendor are high technical standards and the capability to deal with future technological innovations. The vendor must not be in a position to breach corporate vendee secrets.

An assessment of the management philosophy of the vendor is closely linked to the company's quality control policy and may be difficult to obtain. Person-to-person contacts will eventually reveal knowledge, capabilities, personalities, and quality understanding.

An audit of vendor quality control may be requested by the purchasing department to reinforce its initial selection of a vendor. Depending on the nature of the materials under consideration, an initial audit may be extremely detailed or somewhat informal. In either case, the following should be included:

1. The quality philosophy of the vendor.
2. Quality records of other (noncompetitor) vendees.
3. Corporate history and current trends in vendor's industry.
4. Complete product list.

5. Details on equipment, processes, and production capabilities.
6. Vendor's quality assurance system, including quality education and implementation programs.
7. Vendor's controls on procurement of raw materials and subcontracting.
8. Sanitation practices

Maintaining a good relationship with vendors once they are on-line is a necessary and frequently rewarding effort. The purchasing department maintains a relationship that may have objectives different from those of the quality department. The processor's quality records of materials acceptance from each major supplier should be routinely analyzed for trends. Nonroutine audits conducted at the vendor's plant will occasionally disclose practices or conditions which should be brought to the attention of both the vendor and the processor's purchasing department. If needed, advice and assistance should be freely given to the supplier to eliminate problem areas and further solidify relationships. If indicated, interlaboratory calibration tests might be run to show an interest in assisting the vendor. A system of rewarding suppliers for implementing and maintaining process controls and quality systems will promote effective defect prevention and continued satisfactory quality. Because ours is a mobile society, it is possible that a good vendor–vendee relationship might be temporarily lost through loss of key people from either company. The only choice is to start over again, building up the confidence and interchange of ideas necessary to keep the system operating.

Relationships can be improved in unexpected ways. Suppose a vendor's quality manager has been unable to convince his CEO of the need for updated laboratory equipment. A word or two from the vendee quality manager to the CEO that his quality laboratory would benefit from certain modernized equipment, and chances are the vendor and vendee quality managers become long-term allies.

The terms *vendor* and *vendee* in the preceding discussion have been used generically. The impression is that they refer only to suppliers and purchasers of materials and products. However, the same principles apply equally to the relationships between processors and co-packers or processors and warehouses. The co-packers provide an extension to the production lines as a service and are subject to the same scrutiny and assistance that would normally be provided within the processor's plant. By the same token, outside warehouses provide a service which, unless monitored, can adversely affect quality. Specifically, stock rotation or inadequate storage temperature controls may both adversely affect shelf life. Sanitation, pest control, and control of damaged goods are equally critical quality considerations.

The effect of a just-in-time (JIT) contract with a vendor raises special quality considerations. Briefly, a JIT system of obtaining raw materials is based on the belief that no inventory of raw materials need be held at the processor's plant if the vendor can supply the correct material *just in time* to meet the day's production schedule. The system is far from new—the canning industry has used it for years. Field crops such as corn, peas, tomatoes, or peaches are delivered on a scheduled basis to the canning, dehydrating, or freezing plant within minutes of their actual processing. In spite of the complications that arise from tight scheduling, the system works well and permits processing crops at the moment of peak flavor. Similarly, for many years, can, plastics, glass, and paper packaging products manufacturers have supplied the day's packaging requirements to food processors starting on the morning needed and continuing throughout the production period as scheduled. The savings in inventory storage costs and warehouse space needs at the processing plant are enormous, although, in some instances, these needs might simply be moved back to the vendor's manufacturing plants where *they* generate the costs to be either absorbed or passed along.

Quality control in a JIT system becomes critical. There is no time to reject and replace a lot of defective incoming packaging materials. They must be produced right the first time. A properly functioning quality control system must be in place at the vendor's manufacturing facilities before a JIT program is started. A contractual arrangement between vendor and vendee on the disposition of costs associated with substandard shipments is not uncommon, and there are many options available. A shipment of substandard cans might be 100% inspected by the processor, with the added costs absorbed by the vendor; or the cans might be run through the production line and the filled product sorted at the end of the line, with costs of inspection, scrap and rework passed back to the vendor. A postmanufacturing sampling inspection at the vendor's plant prior to shipment might be conducted—by the vendor, the processor, or both—before shipment is approved, thus eliminating the expense of shipping defective materials. In reality, both the vendor and the processor generally maintain a small emergency supply of packaging materials to tide them over in the event of a major quality problem. When a substandard shipment manages to escape the quality control system, it is imperative that the cause of failure be immediately determined and corrected so that it can never happen again.

A word or two concerning single source of supply is in order. There has been growing interest in this concept over the years. The justifications are that better control of quality and costs can be effected by the close relationship developed between the single source and purchaser; increased volume sales from the single source should provide quantity discounts; fewer liaison personnel are required from purchasing, engineering, quality control, and others; and a sense of "family loyalty" develops between the two organizations, with better understanding of mutual needs.

On the other hand, the familiarity and friendships that develop might lead to occasional difficulty in enforcing unpleasant decisions, such as refusing a shipment on the basis of quality or late delivery. Other dangers inherent in using single source include labor disputes, inclement weather, loss of key personnel, power failures, and transportation problems—all of which might interfere with continuous on-time delivery of raw material to the processor. Then there is the ever-present threat of company takeover, which could spell inconvenience or worse to either of the two companies. If the vendor is bought out, the processor might find itself without a ready source of raw material; if the processor is bought out, the vendor might find itself without a customer. The vendor is in the uncomfortable position of depending solely on one major customer, and probably has a greatly reduced sales force (none at all may be needed under the arrangement). If the vendor is dropped by the processor for any reason, there may not be sufficient time for the vendor to rebuild its sales force and remain in business. Perhaps these problems could be solved with contractual agreements, but they are problems nonetheless.

PACKAGING SUPPLIER QUALITY ASSURANCE

Although the following discussion pertains to purchasing of packaging materials, the same principles can be applied to any raw material supplied by an outside vendor. The buyer–seller relationship of the past has occasionally been antagonistic. Worse still, the "friendly arrangements" between purchasing managers and some of their vendors were occasionally responsible for the questionable supplies arriving at the receiving dock. Although these battles and special dealings may never be completely eliminated, there has been a growing trend toward a relationship best defined as "partnering" between supplier and user. It is becoming increasingly clear that the seller, the purchaser, and the final product user have all benefited from this honest and open interaction.

An autonomous purchasing manager for packaging materials developed a quality assurance program with suppliers, which borders on a vendor certification program. Although the procedures put forth are not necessarily recommended, they have become fairly common in companies of all sizes, and in many cases are successful.

Contacts with the vendor's salespeople have been handled generally by the purchasing department. This avoids the possibility of creating violations of contractual agreements which might not be known by, for example, quality control, engineering, or production. On the other hand, frequent meetings of the purchasing department with members of all other groups within his company are mandatory, so that improvements, suggestions, and problems can be aired. These may be identified as formal teams, with a rigorous schedule of carefully structured meetings, or they might be more informal.

Packaging materials are often as important to the quality of food products as are the other raw materials, and for some food manufacturers, the cost of packaging materials is second only to labor costs. Substandard packaging material may result in lost production time, loss of materials, claims, and customer dissatisfaction. In the past, the practice for preventing nonconforming materials from reaching the customer consisted of receiving inspection and tests, with the hope that any defective material would be detected before reaching the production line. Manufacturers have become increasingly aware of the weakness of this system. It is rarely economically possible to sample every lot of every packaging material received fast enough and reliably enough to prevent operational problems. Furthermore, as production facilities become increasingly automated, there are fewer operators on the line to detect substandard packaging materials which went undetected at receiving inspection.

The simple solution is to expect improved processes at the vendor facilities. The first step in this conversion of policy is the preparation of complete, accurate specifications so that the vendor has a clear understanding of the nature of "defective material." The vendor now knows what quality control testing is required for his customer and should be able to provide satisfactory shipments with supporting data. In effect, this shifts the inspection and approval of packaging material from the processor's receiving dock back to the supplier. Certification papers (if required) include production data, size and frequency of samples, test results, and signed reports. Depending on experience with the vendor and the nature of the operation, some companies prefer advanced certification; others might accept certification with the shipment. Some of the responsibility continues to rest with the purchaser, and all shipments should still be subject to audit by them.

In order for specifications to be workable, they should clearly list minimum and maximum measurements, as well as targets. They should be worked out through cooperation with vendors and purchaser jointly. In some instances, the vendor's process capability charts will provide realistic data which both parties find reasonable. Most packaging specifications can be constructed as fast as they can be typed into a computer. However, there are instances where critical performance or intricate design will call for patience and understanding between the two parties. The engineering drawing of a pour spout, for example, may appear to be precisely defined, but tiny deviations from the dimensions shown might product catastrophic leakage. The vendor might provide modifications or compromises which can make the design realistic.

Some of the tests run routinely on paper and fiberboard packaging materials to avoid problems on the line are material composition, inside and outside dimensions, tensile, tear, burst, weight, moisture, smoothness, caliper, moisture, and brightness (TAPPI: Technical Association of the Pulp and Paper Industry). In addition there are hundreds of specific tests for composite cans (Composite Can and Tube Institute), flexible films (National Flexible Packaging Association), cans, glass, paper and fiber, molded plastics, and so on.

Management expects that most directives to install a new company system will be complied with fairly quickly and that the new system will be essentially complete when introduced. For example, a system to prevent multiple disbursements of payables in the accounting department might require several months of planning but will usually be effective the day it is implemented. Similarly, the introduction of new systems to eliminate incorrect time card punch-ins, or pilferage on the shipping dock, or overblown claims in advertising, or excessive absenteeism—these and other problem areas are normally corrected within a few months. Not so with quality control systems. Here, the enormous number of factors which must be right and the even more enormous number of factors which cannot be permitted may take three or four years to uncover and specify. For example, a packaging converter may mistakenly manufacture a double-bulk bag (one inside another) as few times as once every 100,000 bags. This could represent as few as one defective bag in a month's production at the purchaser's plant but could result in several hours of downtime to clean up spilled product and jammed equipment. With a defect this rare, it is unlikely that, at the outset, the packaging specification would initially include a tolerance for double bags, but after a year or two's experience with this problem, it would no doubt be added to the specification.

One would not expect to have the weight of glue on the side seam of a folding carton included on the initial packaging specification. This could be recognized as a critical measurement for the purchaser but would be a difficult characteristic for him to measure. On the other hand, if made aware of the problem, the vendor would be in a position to readily set up a control on the amount of wet glue applied to the cartons during manufacture and could devise a fairly simple test to ensure that no cartons are manufactured without adequate glue.

One of the best sources of information leading to improved specifications is the worker on the line. Given the opportunity, he/she might say one of the following:

1. Every time we run film no. 17 from the supplier in Pittsfield, the right side of the film tears at the package former. This doesn't happen with other suppliers.
2. The carton sleeves all work fine on a new pallet load, but as soon as we get down to the bottom third of the pallet (layer no. 6), they are always deformed, and we have to reject a lot of product.
3. Near the end of each shift, blue ink from the paper printing builds up on the forming tube, and we have to shut down and clean it off. Can't you guys find a better blue ink?
4. Last week we had a skidload of K.D. cases which ran through the shift without a single misfeed. It sure made my job easier. I jotted down the shipment number: ACB69-340.

Most vendors have quality control systems in place. If these programs are not completely satisfactory to the customer in their present form, many customers are now willing, if requested, to assist the supplier in further developing their programs. Some of the requirements which vendors might not normally have in place are documented procedures of their production process, handling and segregation of nonconforming materials, correlation of production samples with specific times of manufacture and with lot numbers of each shipment, and periodic audit samples supplied to customer which duplicate vendor's production samples. On the other hand, for less critical products, perhaps simple checklists of vendor operations could be satisfactory documentation for the purchaser. If, for example, the checklists indicated a number of culling problems throughout a production run, the purchaser would also expect to find a series of inspection reports accompanying any shipment involved, which indicated appropriate culling of defects at the vendor's plant. In the event that this becomes a continuing problem, the vendor would be expected to provide a more formal quality control system. Where informal checklist systems are used, some purchasers find it desirable to receive the checklists in advance of the shipment, as they might consider it necessary to have the vendor conduct further defect removal. Formalized quality control documents, on the other hand, usually accompany shipments.

In examining shipment documentation, it should be apparent that the immense volume of paper could not reasonably be digested by purchasing, quality control, receiving, production, and engineering. As the program evolves, the record flow should indicate a road map where certain documents are directed to specific departments for review before they are collected at some central office, usually in purchasing or quality control. As a guide, the paperwork should be reviewed by the department which needs the information to operate and by an individual who understands how to read and interpret the data. A control chart which shows eight points all in a row with the same value might be interpreted by the untrained reviewer as an example of superb process control. In fact, it might indicate that somebody was falsifying test measurements.

Even under the best quality control systems, it is possible that over a long period of time, as many as three shipments per thousand might possibly be substandard. Here is where the huge data file is most valuable. By careful detective work, it is frequently possible to pinpoint the cause of the reject. In some cases, the problem might not be observed in every portion of the shipment, but might be isolated to the left side of machine #9 during the last half of the second shift on nearly every Tuesday (for example). With detailed supplier data such as this, process improvements may be indicated, and specific products can be culled from the offending shipment rather then outright reject-and-destroy. Costs involved in this inspection are usually negotiated amicably.

The question arises: Who should be responsible for holding a shipment which is accompanied by questionable data? Or, for that matter, who should be responsible for rejecting an incoming shipment when either the data or an audit shows it to be unsuitable? Also, who, if anyone, should have the authority to override a shipment rejection? There is no single answer to these three questions and, yet, they are critical to the control process. Each company finds its own workable solution, depending in part on the function of the departments involved and, to a larger extent, on the abilities of individuals within each company. Where quality control, purchasing, and operations are all equally capable of handling the decision unilaterally, it might be preferable to appoint one department to handle simple, routine rejections or approvals, and to construct a small team consisting of the three departments to act as a material review board. Needless to say, the simpler the better! Obviously, this works best in relatively small organizations.

Carried to an extreme, there is a point at which too much record keeping can evolve from a relatively simple system of reporting. If the vendor is a paper converter, should he also supply documentation from all of *his* suppliers? The paper mill would normally provide documentation with each shipment; the other raw material companies might supply theirs: ink, staple wire, coatings, glues, metal stampings, and so on. Perhaps a statement from the converter which indicates that he has inspected all of these materials and found them within their specifications should suffice. There may be instances where the customer might wish for copies of out-of-specification sample test results, along with the specific causes, and information on steps taken by the vendor to correct the process.

Occasionally, the customer will find the need to modify an existing specification or perhaps create an entirely new specification to accommodate a process, product, or packaging change. These changes require close cooperation between vendor and customer well in advance of the contemplated production. Agreement should be reached on quality control basics: sampling location, size, frequency, methods of test, and reporting. Advance samples from the vendor should be subject to both laboratory audit and production pretest well in advance of the scheduled full production date.

Close relationships with the vendors processes and quality systems can lead to an occasional ethical dilemma. If a vendor should experience a quality problem which requires considerable time and expense to correct, does the purchaser have the right to pass this information on to competitive vendors? Although this might appear to be unfair, there is a workable solution that some manufacturers have found to be acceptable. During the early discussions of closer working relationships between vendor and purchaser, this subject of sharing proprietary information with other vendors needs to be explored in detail. Most vendors will accept the idea, with the understanding that by sharing their own efforts with other vendors, they may have as much to gain in return. If a vendor wishes to exempt specific areas of trade secrets from this agreement, it should probably be expressed in writing.

The "partnering" concept of the vendor–purchaser relationship tends to lead toward the single-source principle. As a suppler and the purchaser grow closer together over time, understanding needs and capabilities, the purchaser might come to treat the supplier as part of his company—a partner. With this close interchange, the purchaser might well decide that there is no need for a second or third source. On the other hand, there is no one best way to manufacture corrugated shipping cases. Each supplier of corrugated has developed his own production techniques with his own specialized equipment and has, over the years, discovered solutions to countless production problems, many of which are unknown to his competitors. This is one of the compelling reasons why so many purchasing managers prefer two or three sources.

Record Keeping

Quality documents pertaining to shipments should be accumulated by either purchase order number or shipment number. In addition, vendors should supply calibration information related to their test equipment and reagents. By using available computer tools, the paperwork can be kept to a minimum. For example, scanner outputs, fax-modems, floppy disks, and CDs containing data can be transferred electronically to a central filing system with minimum effort and keyed to shipment numbers.

The use of specialized universal product coding, in addition to the product UPC information which appears on shipping cases (for example), can be tailored to fit specific needs of both vendor and purchaser.

The major contribution to the partnership concept of quality control between the packaging material vendor and purchaser was expected to be a combination of reduction in defects on the processor's production line as well as lower costs of purchased materials by elimination of manufacturing interruptions at the vendor's plants. A greater benefit has been the ability of the processor to adopt a "Just-in-Time" delivery schedule for purchased materials. As the defect level of incoming materials disappears, it is no longer necessary to maintain high levels of backup stock on the warehouse floor. As one pundit described this change: We have replaced "Just-in-Case" production with "Just-in-Time." Eliminating the need for this cushion of materials also frees up a significant amount of capital. Obviously, for those with uncertain delivery operations resulting from occasional severe weather, unpredictable harvest conditions, unscheduled labor conflicts, extremely long supply routes or other traffic problems, "Just-in-Time" cannot be taken too literally. Allowances must be made for these extraordinary circumstances.

A note of caution: This approach appears to depend on one purchasing manager and one supplier sales representative. Although this system can be quite effective, the danger of losing either of these individuals through promotion or otherwise might severely handicap the business relationship and effectiveness.

SUPPLIER CERTIFICATION PROGRAMS

In much of the above discussion, the word "certified" has been used to describe a guarantee of product quality or a recognition of a supplier's quality qualifications. The USDA has offered a certifying service for many years. Since the 1970s, the United States Department of Agriculture has furnished an inspection service which provides certification of food processing plants and the quality of processed fruits and vegetables, honey, molasses, nuts, sugar, syrups, tea, coffee, spices, and condiments. The regulation is found in USDA CFR TITLE 7, Chapter 38 Food Safety and Quality Service, Part 2852. Plant certification requires a plant survey by the USDA to determine whether the plant and methods used are suitable in accordance with their regulations, as well as those of the FDA Good Manufacturing Practice in manufacturing, processing, packing, or holding. If the plant meets the requirements of the survey, product inspection services may be performed. Certificates of sampling and loading are issued. Official grade stamps are affixed to those lots packed under either continuous inspection of the USDA or packed under the Quality Assurance Program of the USDA. Additional details of the regulations, including sampling plans and acceptance levels have been published in a manual titled *Regulations Governing Inspection and Certification of Processed Fruits and Vegetables and Related Products* (USDA, Food Safety and

Quality Service, Fruit and Vegetable Quality Division, Processed Products Branch, Publication 0-310-944/FSQS-378).

Many companies have customized certification procedures as part of their total quality management (TQM), in contrast to the packaging purchasing manager's example discussed above. These programs appear under such other titles as Supplier Improvement, Vendor Partnership, Supplier Qualification, Preferred Supplier, Supplier Development, Vendor Excellence, Supplier Quality Management, Select Supplier. By spreading the responsibilities and goals over many company departments, the shortcomings of the packaging materials example above are reduced. Some food companies will certify suppliers without audit provided that they have been inspected by recognized external organizations, such as the American Institute of Baking, or by qualified food industry consultants, or have become ISO 9000 certified. This is certainly a shortcut to certification, as large companies have reported that it can require three to five years to achieve fully qualified approval. Some of the vendor's major quality control areas which are examined by various companies include the following

Continuous improvement	Process control
Data management	QC audit system
Experimental design techniques	Recall procedures
Fault tree analysis	Record retention system
Good Manufacturing Practice	Sampling procedures
Hazard Analysis and Critical	Sample retention system
Control Points	Sanitation—macro and micro
Kosher certification	Specifications
Laboratory capabilities	Test methods—accuracy and
On-time delivery system	precision
Contamination control	Traceability
Process capability studies	Training programs

In the simplest form of a supplier certification program, the quality control manager prepares a questionnaire for the potential vendor, listing titles of quality control procedures considered essential to providing satisfactory materials to his company. The questionnaire is filled out by the vendor and returned to the company's purchasing department where a decision is made to "certify" whether the vendor is acceptable.

Another simplified system requires that each vendor's shipment be accompanied by a quality certificate, with or without supporting test data and control charts demonstrating compliance with specifications. These certified shipments may be audited by the purchaser's quality control department.

Although these simplified methods appear somewhat elementary, they may be satisfactory where the products involved are ordinary and uncomplicated. In some cases, they may also be all that a potential supplier will tolerate from a small lot buyer.

At an intermediate level of program sophistication, some customers will conduct plant visits to rate potential suppliers on several specific criteria before certifying them: statistical quality control procedures, product quality, process capability, service, and price. Selecting the initial candidates for certification is usually based on the dollar value, critical nature of the product, volume of material supplied, or perhaps single-source suppliers. Generic product suppliers (salt, oils, sugar, corn syrup, simple maintenance parts for machinery) can be included after the critical product suppliers are in the program. These plant visits can be most fruitful if the supplier receives a clear understanding of how his product is to be used. They may provide suggestions to improve the use and decrease the cost of their product.

Larger companies have developed more detailed programs. More than one company has formed a team to visit the supplier's plant to directly review the routine quality control procedures as well as the quality control tools in use for troubleshooting and product improvement. The team is composed of members from purchasing, production, and quality control, and starts its investigation with the vendor's top management. If indicated, the team may request that he attend a seminar on total quality management. Obviously, this last step must be accomplished with much tact and good judgment.

The common goal of the certified supplier programs is the assurance of receiving problem-free materials or services on time, all the time. Some companies will contribute to this goal by providing technical assistance to the supplier for installing quality control systems. A second common goal is to reduce cost. This can be accomplished by reducing losses from defective materials, from eliminating receiving inspection and returned shipments, and from supplier process improvements, possibly aided in part by the purchaser's technical personnel. A third goal is to improve the quality level of the product purchased through improved operations by the supplier. A fourth goal expressed by a few companies is to develop a virtual partnership between vendor and purchaser to their mutual advantage. With the exception of the USDA program mentioned above, purchasers will generally certify specific products, not the vendor's complete line of products or entire operation.

The most complete formal programs, generally in large companies, are sometimes considered proprietary, and nondisclosure agreements must be signed. These programs generally include stages of certification which demonstrate that a supplier has furnished product or service which

Consistently reaches an established quality level
Complies with all governmental regulations
Meets Just-in-Time requirements
Continually improves product quality
Improves price/value relationship

Approvals generally are accomplished in three to five stages (or more), and suppliers may be identified by a number of classifications, as shown in the following:

Stage	Classification	Suggested descriptions
I	Potential Restricted Provisional	Might become a supplier after product or service has been sampled and reviewed
II	Limited Conditional Authorized	May furnish products on limited basis, subject to receiving inspection
III	Acceptable Approved	Preceded by plant visit Certificate of analysis required
IV	Accredited Qualified Preferred	Selected to participate in certification process Delivers consistent, excellent quality
V	Select Certified	Meets all of the TQM process, product, and service requirements
VI	Continuous	Periodic audits of plant and product

As complex as these certification programs appear to be, there is increasing evidence that they are effective in both cost savings and quality improvements. It is expected that this business technique will continue to grow in popularity.

14 Implementing a Quality Control Program

MANAGEMENT COMMITMENT

When top management directs that a quality program be created and adhered to, the quality manager's job is relatively straightforward. He arranges for staff and sets about organizing a system starting with specifications, manual, methods of sampling, testing, reporting, auditing, methods improvement, cost reduction, and training. The first efforts are usually centered around production.

Assuming that the initial efforts are successful and that the directive is still in force, the next logical step is to install quality control programs in other departments: shipping, purchasing, accounting, sales, marketing, personnel, engineering, and so on. Chances are that this is not exactly what top management had in mind. Until the 1940s, quality control meant simply to "inspect the raw materials and the finished product, and don't let anything bad get by." There are still organizations with this simplistic attitude. There may even be some where the product and process are so simple and the employees so dedicated to high-quality production that it may work satisfactorily. For some older food companies, the main goal for years has been to get as many cases of product as possible out the door. Using quality control to help production attain this goal would be acceptable to company personnel, but quality control should not be allowed to "waste their time" in other departments.

The concept of total quality control—that is, quality control encompassing all of the company's functions—is still fairly new. It is often difficult to establish because of the "closely guarded turf" syndrome: "I'm the accounts payable department manager, and nobody from quality control or anywhere else is going to tell me how to run my business." Yet, the quality of work in the accounts payable department may be very poor. Payments may be so slow that thousands of dollars of prompt-pay discounts are lost; $X\%$ of checks may be sent out with the wrong amount, to the wrong account, or in duplicate. $Y\%$ of invoices may be misfiled or destroyed. $Z\%$ of invoices may be for unsatisfactory services or services not performed at all. Certainly, the quality control principles using sampling plans and statistical analysis can be used to pinpoint weaknesses in the system and provide measuring devices for process improvements. It would not be unexpected for the accounts payable manager to tell the quality control manager, "Don't try to educate me with your quality control principles until you've solved all the problems of poor product quality control on the production line. That's what you were hired for." In many cases, he could be right.

With this in mind, we restrict our discussion of implementing a quality control system to those areas of a food processing company directly involved with operations.

GETTING STARTED

To begin on an optimistic note, a statistical quality control program can be created in a company with or without the direct support of top management. Support merely makes the job easier. If management has selected a philosophy—such as statistical or motivational—the first step has already been taken. From there on, every step should refer to that philosophy.

It is possible to buy books, videotapes, and consulting services that supply prepackaged quality control programs prepared for other companies. The principles might be applicable, but the details may not fit into the product safety, cultural atmosphere, product line, governmental regulations, and physical processing equipment of every company. Trying to educate managers in a darkened room with videotapes aimed at another type of industry will likely put some of them to sleep. A program aimed at certain hardware industries may be built around such principles as *reliability, mean time between failure,* and *warranty control,* much of which does not apply to food products. The quality control program of a food product cannot be based on a system that even suggests that a food's reliability makes it safe to eat 87% of the time, or that, if it should cause sickness, it can be returned for repair under a warranty certificate. Some programs have zero defects as a goal. In reality, the only way to reach zero defects is to widen the specifications. As soon as the line worker realizes this, the program loses its punch. This does not mean that there is no value in a goal of zero defects. It is an admirable way to present progress charts. Where attribute acceptance programs are used, an improvement in quality control which permits dropping the acceptable quality level (AQL) to the next more stringent table—closer to zero—should be heralded with loud applause throughout a company bent on improving the quality level.

Avoiding blind acceptance of hardware or electronic quality control programs cannot be emphasized enough. In an article published in *Quality Magazine* (Heckel, 1988), a production manager of Hewlett-Packard Co., the world's leading test and measuring equipment manufacturer, admitted to a major problem with product quality. The article describes how, in 1981, there was a 50% chance of product failure resulting in the equipment being returned for warranty repairs within certain areas of the company. By instituting a stringent new quality control program, the annualized failure rate was reduced stepwise to 14% by 1985. After a few swings in either direction, the product failure rate leveled off at 4% by 1989. The target failure rate for the 1990s was estimated to reach about 3% to 4%.

Certainly, a quality control program such as this one, which has been claimed a huge success for a company in the electronics industry, cannot be used in the food industry without major modifications. In the first place, it is inconceivable that a food product with a 50% failure rate would ever be considered ready to market. Furthermore, the concept of repair under warranty does not exist in the food industry. Finally, the thought of a 3% failure rate as a goal for a food product is ridiculous. If 3% of carbonated soft drink cans were failures (leaking cans permitting the carbon dioxide to escape, pull tab openers that fell off when attempting to open the can, product contaminated with cleaning fluids, jagged edges on the pour opening, moldy product, wrong flavor, missing formula ingredients such as acid, color or flavoring), then 1 of every 17 six-pack cases sold would result in a dissatisfied customer. This is totally unacceptable for any food product.

Some programs are aimed at quality control personnel and tend to be highly mathematical. If presented in digestible doses, they may quickly and efficiently train those personnel. On the other hand, a mathematics-based program, loaded with statistical terms and dull tables, will not be readily embraced by most managers or supervisors whose backgrounds make this material difficult to absorb. The lesson to be learned here is that any potential program should be screened before presentation; if it is not applicable to the company, another program must be found or created.

AN IN-HOUSE PROGRAM

A major food processor tried a formal mathematical program and found that it was poorly received. The company then set up small team meetings where operations and line personnel discussed cost-savings projects and where fragments of statistical training programs were taught. This approach worked. With specific company goals to keep their interest, the teams almost immediately introduced methods modifications for product improvement and cost savings. At the same time, they were taught the meaning of statistical control in a relatively painless fashion, so that they were familiar with X-bar and R charts when they encountered them on the plant floor. Efforts of this type must be closely monitored because they generate costs in lost time, labor, and production.

These teams were not modeled after Japanese quality circles. They consisted of management, supervisory, and hourly personnel, and the discussions were guided along specific problems. This gave everybody an opportunity for input without wasting time talking about unsound engineering approaches to complicated systems. To start a team in the right direction and to show how effective team efforts are, a simple demonstration was conducted. A team leader went through the process of preparing a peanut butter sandwich, but without plans, tools, or materials readily available. He crisscrossed the front of the room looking first for the bread, then for a plate, then a knife, and then the peanut butter. He stopped to perform other actions that had no bearing on the project, such as getting a drink of water, stopping to open up a pack of chewing gum, and looking for a wastebasket for the wrapper. After 10 minutes of aimless and uncoordinated motions, the peanut butter sandwich was finished. A single member of the team was then asked to list all the wrong actions and suggest correct methods for performing this simple task efficiently. A team discussion followed to demonstrate how many additional and better ideas could be generated by group contributions. After a few days of practice with actual production line situations, the team came up with a few minor but productive suggestions relating to their jobs. Team members gradually realized that their input was being used, and employees at all levels were gratified to be considered a part of the production team. As the team became aware of the positive results, some technical and statistical information from the Juran and Deming principles was slowly introduced. In this setting, people were more receptive to statistics, and the education program was underway.

Next came the introduction of outside professional training. Some employees were sent to community college statistical quality control courses, and a private consultant was brought in to teach courses to groups of employees who were now partly pretrained and ready for the heavier material. Several quality control employees were also selected by the consultant for specialized 5-day training in process control courses. These specially trained personnel then presented 4-hour classes each day for 3 days to the plant production personnel. Rather than presenting detailed procedures, these courses included an overview of sound statistical principles, covering much of the same subject matter as this book, but in a somewhat theoretical manner: sampling, distributions, control charts for variables and attributes, data collection, and chart analysis. This endeavor turned out to be only partially successful. It was too general and theoretical and failed to give the plant people the specific tools they needed. Management considered this part of the program as "planting seeds" of statistical quality control; if enough seeds were planted, they reasoned, sooner or later something would sprout.

The production and plant managers were invited to the consultant lectures as well, but they chose not to attend. This was unfortunate, because the hourly plant people read this not only as a lack of interest but as an indication that the managers probably did not believe in the idea of quality control. Unless workers are assured that their manager thinks the program is important, they will not pay much attention at the classes.

One of the first statistical tools presented by this company to line workers was Pareto analysis, offered at first as a dollar-saving tool. During half-hour sessions one day per week, the team was taught this technique, and eventually how to apply it. Pareto analysis is simple to present and understand, and it was shown to be powerful.

The company which tried this approach is a large food processor with plants nationwide. When the program started, six teams were assembled; due to their successes, the number was expanded to over 100 within four years. These 100 teams were credited with generating over 7 million dollars in savings per year. Although the program was originally under the direction of the quality control department, it grew to the point where it had to be transferred to a separate, ongoing operations department function.

This company believes it still has a long way to go before the training program can be considered a complete success, but progress has been steady; after four years, the spirit of cooperation continues to spread through all departments. Some supervisors still greet suggestions from the quality control technician with, "I'm busy. Stop bringing your problems to me." Pressure applied indirectly—by the quality control manager to the plant manager and down through the various layers to the uncooperative supervisor—usually results in grudging compliance. In time, this supervisor will realize that the suggestions are valuable in improving his department's productivity; eventually, cooperation will be easier to obtain.

This particular company once depended on a suggestion program to generate new cost-saving ideas. Employees who contributed successful ideas were compensated for them. That program has been all but abandoned as unsuccessful, however. Its most serious problem was its bureaucratic nature: a suggestion could take months to work its way up through the levels of management and, if approved, more months back down again before any action could be taken. Adding to the failure of the system was the fact that some approved ideas—for which employees were rewarded—were never implemented. The new team programs have a compensation and awards plan built in. Successful teams receive cash awards at the end of the year, calculated as a percent of the savings actually generated. Interim awards such as gifts, jackets, and pen sets are presented at special dinners or sports events, with the presentations made by senior members of the firm, usually vice presidents, to impress upon the teams their importance to the company.

TEAM QUALITY SYSTEMS

The preceding examples of team development and operation are each based on relatively unstructured beginnings, evolving into formal team programs after their usefulness has been demonstrated. Over the years, universities and consulting firms have refined systems for team structure, training and operation which have had successful implementations. It is unlikely that agreement will be reached on the "one best method" of introducing teams because of the wide differences in managements, culture, and diversity within industries.

Team activities are performed in a wide variety of ways in food companies. Although often thought of as modified quality circles, the concept is not new. In November 1932 McCormick and Company established a team known as the "Junior Board." They met periodically to discuss improvements in operations, and reported their suggestions to the senior board for action at their next meeting. During the first five years of their operation, over 2000 suggestions were accepted and implemented, with resulting improvements in plant production estimated at 30%. The specific methods of composition, operation, and compensation have changed over the years, but the overall concept remains.

Articles in food trade magazines and newspapers report a steady stream of successful team projects in a wide variety of areas: new production techniques, refining accounting procedures, moving facilities to less expensive locations, absenteeism, accident reduction, processing customers' orders, reduction in line changeover time, defect reduction, material ordering practices, improve sanitation, reduce energy consumption, reducing breakage, customer-driven product specifications, improved stock rotation, community–plant relations, and on-time delivery.

The team approach to improvements and problem-solving may not work for everyone. Some of history's product and system inventive geniuses probably worked best without a team of "equals." Thomas Edison, Albert Einstein, and Luther Burbank are examples. Where a company is fortunate enough to employ some near-geniuses, perhaps it would be best to allow them to work alone, if that is their expressed need. And there are many formalized team organizations directed by an autocratic coach, who cannot (and perhaps should not) encourage or accept empowerment of their individual members: sports teams such as football, baseball, basketball, soccer, and hockey come to mind. The same is true of symphony orchestras, where the will of the conductor may not be challenged without destroying the unified perfection of the ensemble.

A quality improvement team cannot be expected to replace a quality control system. On the other hand, it is unrealistic to expect any single employee to have the ability to solve all quality and productivity problems and provide a flow of improved quality and productivity ideas. Thoughtfully selected and trained employee teams can contribute to improvement in every company's operation and profitability. Note the wide variations in quality team operations:

Team sizes.   These vary widely according to complexity of the project and the needs perceived by management—there are no rules for minimums, but it is wise to limit the size to avoid interminable discussions without conclusions. (A major candy company found an exception to this limitation: at one point in the development of a new candy product, they claimed to have had 150 people working on the project.)

Leaders.   It is possible to have successful teams without leaders, but chances for direct action are improved with the appointment of a leader, either by management appointment or by team election. Some teams seem to work well with revolving leadership.

Team assignments.   For greatest opportunities, these are often presented in general terms to allow for open-minded exploration of the entire subject by the team. More frequently than not, the final problem statement turns out to be quite different than originally thought by management and may result in more far-reaching quality improvements and cost-saving results.

Rewards.   There is little agreement on the methods for rewarding teams for their successes, in part because of the possibility of fomenting team jealousies, and partly because of perceived uneven-handed recognition of groups over individual efforts. Agreement by labor unions to a reward plan has been a problem for some. The Human Resources Department is faced with a decision on whether to reward individuals as determined by classical evaluation methods, or to review the individual's performance in group activities.

Schedules. Different teams within the same company might have varying meeting schedules depending on complexity of the project, technical ability of the members, and availability of both company and outside information. In addition, the urgency of the project will influence the schedule.

It would appear from this discussion that quality improvement teams are all unique, but successful teams do have six things in common: management support, training in quality team techniques, a stated mission, interdepartmental membership, budgeted funds, and empowerment.

Just as quality improvement teams require input from various departments, the Research and Development Department's efforts eventually involve other departments before introduction of a new product. Many companies have formalized the procedure by creating cross-functional R&D teams. Initial ideas for product research may originate in the marketing department, or from a customer, or from any employee's interesting idea. An R&D team will receive input from marketing regarding pricing, packaging, portion size, and other product specifics. Production, engineering, and quality control provide information regarding specifications, test requirements, and equipment and procedure needs. Purchasing locates and prices ingredients and packaging materials. Cost accounting calculates prototype standard costs, and determines pricing. Outside sources of information may come from equipment manufacturers, customers, raw material suppliers, and the advertising agency.

There are many sources of detailed information on various types of team operations. Figure 14-1 illustrates four concepts of team formation. Type IV is an informal and ineffective choice, selected by management which is not particularly serious about the use of teams. Type III is a step in the right direction, where a manager realizes his shortcomings and solicits assistance from other departments to act as an informal team. Type II is a fairly effective method of team formation and has been used to solve specific production or quality problems. Type I teams have the greatest potential for continued success.

Some authorities have segregated teams into various classifications, depending on their longevity and scope. Examples are (1) the project team, which is assembled with a single specific goal in mind, and usually consists of a small group of employees who are all familiar with the problem or system to be explored; (2) the permanent team, generally appointed by management to meet on a prescribed schedule and to consider opportunities within a specific area or group of areas; and (3) the self-directed work team, which is empowered to consider any and all aspects of an organization's processes.

Once a plan has been formalized for team selection, the next step is training. For some quality improvement teams, it may be advantageous to teach the basics of control charts and elementary statistics; for others, this need can best be performed by experts in the company or contracted from outside consulting services. In addition to providing guidance on principles of cooperation and ground rules

Type I Management forms
a team which represents
every department. The
goals are problem solving,
process and quality
improvement. Training is
furnished to all.

Type II A team is formed from several
experienced workers in the affected
departments who are directly involved
with a specific problem. They are likely
to cooperate in making it work, since
they have have a personal commitment
to suggested solutions.

Type III A department head has no training or
experience in the financial or engineering skills (for
example) required to solve a quality improvement
problem. He assembles members of the financial
and engineering departments to form a team which
should fill in the voids of his qualifications.

Type IV Nobody in management has a solution to a new quality
problem. A team of management friends meets to look for useful
ideas.

Figure 14-1. Team formation concepts.

for conducting work sessions, a number of techniques need to be taught to the
members:

Brainstorming
Process flow diagramming
Fishbone chart construction and analysis
Pareto analysis
Scatter diagram preparation and interpretation

Finally, the Deming cycle: PLAN / DO / CHECK / ACT (also referred to as the
TQC improvement cycle, or the PDCA cycle) must be understood by all members
so that the proposals of the team can be finalized. Team recommendations in the
food industry may require an extra step to assure that the end product complies
with governmental regulations and safety standards.

STEPWISE PROCEDURES FOR TEAM PROBLEM-SOLVING

After the initial training programs are well underway, each team should be presented with a practical problem in its own production area. Rather than starting

with a major problem, the project should be broken down into smaller projects which can be more easily handled. To facilitate understanding, each project is shown to the team as a fishbone diagram. Causes are pinpointed through team brainstorming and placed on the diagram. The next logical step is to analyze the importance of each potential cause by use of a Pareto analysis. It makes sense to start with a numerical Pareto analysis and then follow it with a dollar analysis, because the most frequently appearing problem may not be the most costly. This now gives the team a series of project priorities based on the value to the company.

Once the major problem has been selected, the process is studied in detail. The tools generally used for this are process flow charts and process flow diagrams. At this point, the team is ready to brainstorm for possible solutions. Rules for brainstorming are simple: A facilitator or leader is chosen to maintain order, the facilitator asks for suggestions at each point on the fishbone diagram, and the team offers brainstorm thoughts one member at a time. All thoughts are written on a blackboard. At this point, the team proceeds to data collection.

The three main reasons to collect valid data are discussed by the team, with respect to the problem at hand:

1. Analyze the process to determine its capabilities and limitations, with the goals of improving the capabilities and eliminating defects.
2. Provide the basis for action to achieve a state of statistical control over the process.
3. Inspect parts or products for acceptance or rejection.

It should be emphasized that collecting data when the solution to a problem is not fully understood should be aimed at *exploring,* not *proving.* Data should be studied thoroughly with all available tools: frequency distributions, histograms, and line charts. At this point, a Quincunx might be introduced to demonstrate the generation of normal distributions through random motions of beads fed first at one target and then at another. This demonstration should lead into an explanation of three-sigma process control. The theory is followed by fitting the data collected on the plant floor to a normal curve and calculating the control limits. Finally, the data are plotted on X-bar and R charts to further simplify discussions of the process. By this time, team members should have a fairly clear concept of problem-solving techniques and the tools of quality control.

The procedures for corrective action are introduced, and, finally, quality control reporting formats are presented and explained. In the company described in the preceding section, the quality control manager prepares daily summary reports which are discussed at staff meetings. The major report is one page of HACCP summaries, pie charts showing defect causes, Pareto charts highlighting major problems, and control charts for plant operations. These reports can be prepared by hand but are more commonly computer-generated. Teams then review quality control recap reports, which are much more detailed and include summaries of all team projects. The recap reports form the basis for management decisions on the scheduling of future team projects, as well as decisions to install new equipment, procedures, or processes implementing team suggestions.

This completes the first cycle of team training—problem discovery, problem analysis, data collection and analysis, proposals for problem solutions, review, and implementation, and then on to another problem discovery and solution.

Example

Let us consider a cost-improvement project in maraschino cherry processing. Historically, a plant had stationed four inspectors at the raw cherry receiving tables to remove blemished, crushed, and spotted cherries. An additional eight inspectors were stationed along the bleach discharge belt to pick out defectives that had passed the first table. The reasoning behind the double-sized crew was that the bleached fruit showed up blemishes more clearly than the red-colored fruit at receiving. Finally, three inspectors were distributed between the color and glass operations.

Operation	Raw	→	Bleach	→	Color	→	Flavor	→	Glass
Cost per pound	$0.40		$0.70		$1.00		$2.00		

One of the improvement teams asked the logical question, "Wouldn't it be more profitable to place most of the inspectors at the beginning of the line, where discarded defects are worth only $0.40 per pound, rather than throwing away defects worth $1.00 a pound after coloring?" A *yes* answer would leave some doubt about the effect this change would have on final quality. No amount of discussion would resolve the question without hard data to support the proposal. A series of tests was conducted with various numbers of inspectors at various locations, using statistically designed experiments. The final results showed that two inspectors were still needed after the color operation, but that only 10 inspectors stationed at the receiving tables could produce satisfactory product quality—thus saving two inspectors and thousands of dollars needlessly spent on bleaching and coloring defective material which would subsequently be scrapped. (The savings, though substantial, were less than expected because markets existed for flawed product, although at substantially lower prices.)

PROGRAMS WITHOUT MANAGEMENT SUPPORT

The program involving cherry processing was blessed with top management support and spearheaded by a dynamic quality control manager who made it work. But what happens to the team effort if this dynamic manager moves on to another company? If he is replaced by an equally competent leader, the program will continue as long as it receives continued support. If, on the other hand, no leader appears, it is possible that each team will become autonomous, and effectiveness is bound to suffer. When everybody is in charge, nobody is responsible, and teams will crumble.

Another ugly possibility is loss of support due to a change of heart by the chief executive—or worse still, a change in chief executives to one with other priorities. Changes are not merely possible; they are likely. Top management does change from time to time, and the possibility of a merger bringing in a new quality philosophy is not unheard of. An equally unpleasant scenario, one which may be fairly common, is the directive by a new chief executive to the quality control manager: "Here's $100,000. Go build a quality control department."

When top support is not available, a new quality control manager might attempt to repair the first quality problem he finds. Such a poorly planned effort could considerably shorten his career as a manager. The trap of moving first and planning later is sometimes referred to as the *ready–fire–aim syndrome*. Without management support, any false move can scuttle the possibilities of a successful program.

Another method of installing a quality control program in a company, with or without management support, is the one proposed by this author and described in Chapter 1 (Hubbard, 1987). First tried by the author in the mid-1940s before there were any guidelines for food industry process control, it has since been successfully introduced in several other companies. In three companies of considerable size, this quality control system was constructed step by step, by a single technician. As the systems generated profits, additional personnel were added and the pace of development picked up.

Starting with Step 1, note the tentative approach of finding an area of quality control in need of assistance using either Pareto or *political* procedures. Why political? When management of one company was shown that eight packaging lines were each losing up to several thousand dollars per week from overfills, they expressed righteous indignation and disbelief that they could be such poor and ignorant managers as to allow this to happen. In such an egotistic atmosphere, it was obvious that another approach was needed. A Pareto analysis revealed excessive, intermittent product dust content in the finished package. This problem was far down the critical quality list from the overfill problem, but it was a condition which aggravated management because of weekly consumer complaints. It was low on the Pareto list because the total value of a year's worth of complaints amounted to less than $1000 worth of product. On the other hand, it presented an opportunity to relieve production management of a constant abrasive conflict with the sales department. After a few weeks of sampling and statistical analysis, it was found that the excessive dust was caused by buildup in the ends of elongated filler hoppers. This showed up in the capability analyses as a process with two sets of dust quality levels. The design was changed to provide pyramidal discharge hopper designs that completely eliminated the intermittent dust deliveries, and the fledgling quality control technician became a minor hero to the production management.

With this success, cooperation was now possible in solving the major quality problem of overfill. Step 2 was to use statistical techniques to calculate the process capability limits of the filling equipment and to calculate the control limits. Step 3 established sampling locations, frequencies, size, and methods of testing and reporting. The product under study was a one-pound can of granular food weighed and sealed at 120 cans per minute. Because of the volume of samples, the sample size was set at triplicates from each filling head on approximately half-hour intervals. A study of process outliers quickly led to the cause: overadjustment of the filling head scales.

Step 4 called for correcting these assignable causes and calculating the improved capability. On a test basis, the assignable cause of variation was temporarily removed by forbidding scale adjustment unless the product weight exceeded the process control limits. As expected, the weight distribution narrowed dramatically, which led to recalculating the new and tighter process capability with a corresponding lowering of the target weight. The remainder of Step 4 required reporting dollars saved and other accomplishments to management. By repeated iteration, overfill on the test line was reduced from the 1/4-ounce overweight to 0.001-pound overweight, with a monthly saving of close to $30,000 worth of product per month.

Step 5 repeated the preceding two steps until no further improvements were apparent. It was during Step 5 that the first real change in management's quality control philosophy was seen. Bear in mind that up to this point, all work was accomplished in a near-hostile environment by a single engineer. However, the success evidenced at Step 4 convinced management that statistical process control really did work (even though the study took nearly four months), and the engineer was rewarded with two new employees. In realistic terms, this meant that the final improvements expected from Step 5 were completed in just a few weeks on the remaining production lines.

Although this quality program states that Step 6 should consist of experiments to modify the process and thus improve productivity, Pareto analysis may reveal more fruitful areas to explore before wringing out all of the increased productivity from this first project. Also, from a practical standpoint, management might be tired of hearing the same success story. There are situations where Step 6 may be skipped temporarily and Step 7 pursued.

Step 7 leads us to another line or another function. It also suggests another department, but it would be well to wait a long time before having the confidence to tackle this very difficult quality control challenge. The chances are good that there are enough glitches in the production area to be dealt with before stepping out into the minefield of other departments.

Finally, Step 8 installs quality attribute acceptance sampling plans where indicated as safeguards for quality in the process and the finished product. A natural expansion of this effort leads to a company-wide audit system. Then—as so many statistical quality control plans demand—it starts all over again at Step 1. It should be easier this time!

TRAINING QUALITY CONTROL TECHNICIANS

Most of our discussions have centered around training nontechnical personnel to use statistical control techniques without teaching them heavy statistical procedures. The language to use with management, we have seen, is dollars saved, profits increased, productivity improved, complaints reduced, and scrap and rework minimized. The language to use with operations personnel is largely related to fewer defects, job simplification, process improvements, as well as cost savings.

Now we are concerned with the language to use in training technical personnel who are required to collect and analyze data and processes and to understand a host of technical terms. These technicians need to learn how to operate chromatographs, pH meters, HPLCs, mass spectrometers, infrared spectrophotometers, and bacteriological and other multisyllable pieces of expensive scientific equipment. They may be asked to explain to line personnel the meaning of control charts posted on the production floor.

Perhaps this problem is not as great as it appears on the surface. If we agree that each person has specific abilities and learning limitations, then the training should proceed in directions that are comprehended and, hopefully, enjoyed. In large companies, quality control technicians may become specialists in one or two technical fields. In small companies, they should be taught "everything," and then supervised closely in those areas which seem difficult to understand. People differ, companies differ, and specialization needs within each company environment differs. As a logical result, the training of quality control personnel cannot be standardized.

An excellent starting place for training is the quality control manual. As pointed out earlier, the manual should cover in detail all of the principles, test techniques, calculating, product specifications, reporting and recording, audit procedures, sampling locations, quantities and frequencies, production flow diagrams with HACCP precautions, sanitation, and so on, and so on. To quote an old Chinese proverb: *"I hear and I forget; I see and I remember: I do and I understand."* There is much wisdom in that quotation. There is one more phrase that might be added: *"I teach and I become expert."*

An effective method to teach a new employee is to pair him with a seasoned inspector who is making rounds, inspections, and tests; the older inspector teaches the newer one to actually perform the functions under close supervision. As the proverb indicates, this adds to the knowledge of both. A safe technique for training a newcomer to run a complex, analytical instrument is to allow him to run (with supervision) tests duplicating those run by experienced employees until he reports identical results.

It is strongly suggested that statistical techniques be taught outside the plant and laboratory. Excellent courses are available at community colleges and through local chapters of quality control and mathematical societies. Seminars are offered by experts in statistical quality control, some of which may be applicable to the food industry. Teaching statistics in-house is certainly possible, but time and personnel might be better utilized on day-to-day quality functions.

SUMMARY

In this chapter we have explored methods of implementing a statistical quality control program, and we have discussed some of the training methods available:

- Use of an outside consultant program with management support and direction can be successful if the program content is compatible with the company's needs. Beware of the "one-program-fits-all" program mill.
- In-house team development takes time but can work well. Management support can be earned through profit-generating project resolutions.
- A universal program started from scratch without the initial support of management was developed for several food companies and has subsequently been enthusiastically adopted, implemented, and endorsed by management.
- A few suggestions for training quality control technicians are offered.

15 The Computer and Process Control

Yesterday the computer was an electronic slide rule with pictures and limitless file folders. Today it can also be programmed to listen to commands, compare inputs, make decisions, and send orders to machines or other computers. Tomorrow—there is no limit! The explosive growth of computer applications has taken at least three major directions, although there is considerable overlapping. One has been total company management through computer integration (CIM); a second has been the development of Artificial Intelligence and Expert Systems; and the third has been the growth of processing automation by computer control.

COMPUTER-INTEGRATED MANAGEMENT

The goal of computer integrated management (CIM) is to optimize processing through the use of linked (integrated) computers covering every stage of a business: design, engineering, human resources, purchasing, production scheduling, quality control, processing, maintenance, materials handling, safety, inventory control, marketing, legal, distribution, finance, and accounting. For some companies, the initial goal of CIM may have been reduction of personnel through automation, but this powerful business tool has turned its focus to improved product quality and process efficiency (cost savings) in every phase of the organization. By integrating the data bases of each segment of the company, instantaneous communication of shared information can optimize the entire operation. In addition to providing real-time information, a CIM system is an excellent tool which can provide rapid answers to "what if" scenarios.

The application of computers to quality control and to processing is a specialized segment of CIM. In theory, quality control techniques can be applied to any function of the corporation. Initially, analysis of production quality by the use of attribute and variable control charts is greatly simplified in a CIM system, because data should be immediately available at every step of the production process. By including data from distribution, purchasing, sales, and accounting, the analysis of process improvement tests can be more realistically evaluated from the overall company viewpoint. In the past, cost information required for a test run, for example, would usually be calculated from estimates of standard costs at each step; with CIM, precise costs should be readily available, eliminating some of the inaccuracies of the older methods.

There are over 500 software companies offering quality control computer programs available off the shelf, making it difficult to select the "right one." Each company offers up to six different programs, making a staggering total of over 1500 programs from which to make a selection. Some of the features offered by these programs are presented in Table 15-1. The seven basic tools are in bold type.

It is suggested that quality departments planning to obtain their first program should give consideration to those which contain the seven basic quality control tools referred to previously. The finer points available in advanced programs can be obtained later, after the basic procedures have been established. There is little need at the outset to have the ability to run analysis of variance, or to have t-test of means available on disk. There is also the danger of blindly selecting a fragment of a program which looks interesting, draws complicated graphs, and expresses results of calculations in complex Greek letters but is an incorrect application for the purpose intended. In one unfortunate instance, an inexperienced quality manager selected a process capability graphing routine from the menu of a new program, because it "looked good." He then incorrectly used it as a process control chart for package fill volume control, causing excessive fluctuations of this measurement in finished product, with accompanying increased production costs.

ARTIFICIAL INTELLIGENCE AND EXPERT SYSTEMS

Some of the early advances in Expert Systems were made in the food processing industry. One company, troubled by the advancing age of their most knowledgeable processing engineer, realized that when he retired, he was taking with him a lifetime of troubleshooting techniques that had proven invaluable over the years. They teamed him with a computer expert, with the goal of transferring his expert production experience to some type of artificially intelligent computer program. After nearly a year of full-time concentration on this project, a series of disks was prepared by which operators could readily find expert assistance at solving production problems with the touch of a computer keyboard. The mind of the expert engineer had been transferred to the computer for ready access! Additionally, this information was expanded with the help of other experts to provide a training program for new employees. Later, another set of expert system disks was prepared for use by the maintenance department machinists. As more information became available, and as process changes required modifications, this program was readily expanded and modified. As most of the information contained in this type of program is proprietary, few detailed procedures for this technique are available in the literature.

Table 15.1. Statistical Software Examples

Acceptable Quality Limit
Accuracy Graphing
Analysis of Variance
Analysis of Means
Attribute Charts
Average Outgoing Quality
Baldridge Award Criteria
Bar Code Access
Bar Code Generator
Benchmarking
Box Plot
Brainstorm Format
Calibration
Cause and Effect Diagram
Check Sheet
Chromatographic Analysis
Complaint Management
Continuous Quality Improvement
Confidence Limits
Contour Plotter
Correlation
Cost of Quality
Criticality Analysis
Matrices
Mean Time Between Failure
Median Charts
Moving Average/Range Charting
Non-normal Distributions
Normality Test
Operating Characteristic Curves
Pareto
Pearson Curve Fit
Probability
Process Capability
Process Flow Diagram
Process Improvement
Quality Manual Format
Quincunx Demo
Real Time Charting
Regression Analysis
Reliability Functions
Run Chart
Sampling Plans

Scatter Diagram
Scrap and Defect Tracking
Short Run Control
Signal/Noise Ratio
Customer Survey Analysis
Cusum
Data Acquisition
Defect Code Management
Defect Cost Calculation
Design of Experiments
Distribution Analysis
Document Control
Failure Mode Effect Analysis
Fault Tree Analysis
Fishbone Charting
Flowcharting
Forms Design
Fourier Transformation
Fractional Factorial Designs
GMP Training and Control
Gage Control
Gage Reliability and Repeatability
Gantt Charting
Histogram
Hypothesis Test
Ingredient Tracking
ISO 9000 Criteria
Simulation
Spectrographic Analysis
Standard Deviation
Statistical Methods
Statistical Process Control
Stratification Analysis
Supplier Quality Assurance
t-Test of Means
Taguchi Analysis
Thermal Analysis
Three-Dimension Plotter
Traceability
Training
Trend Analysis
Variables Charts
Weibull Curve

Note: Basic tools are in bold type.

There is a real need for this type of information. Most experts tend to bury several steps in their reasoning process deep in their subconscious. After the expert has left the company, these reasoning steps may be lost forever, unless recorded as part of an expert system program. A simple example: Once the expert has learned that "left makes loose, right makes tight," he no longer needs to go through this thinking process when he instructs an operator to "turn the controller knob counterclockwise." In addition, the expert is not available for consultation on each shift every day the plant is operating. Sometimes there are situations where two or three experts may be required to offer a solution, and one or more of them might be on a business trip, ill, or on vacation. In their absence, this function is handled in many organizations by the use of operating manuals; but the need to thumb through pages of the "troubleshooting" section to locate the problem description and its cure often takes much too long to be of practical use.

With expert systems available, productivity can be expected to increase by having the latest expertise available at each worker's command, avoiding the necessity to shut down a process while seeking an expert. In addition, by having the best advice on-line, improvement in decisions affecting quality control are achievable.

Following is a simplified example of how a production operator is able to call up an "expert" in seconds to solve a processing problem. Assume that the process requires a precise temperature range for control of the product quality, and the operator observes that the temperature has fallen below the lower limit. He slips his Expert System disk into his workstation computer and observes a message on the screen (Fig. 15-1).

By checking "Temperature," the next window immediately appears (Fig. 15-2). Checking "Too Low" brings up the next window (Fig. 15-3). This window explains the procedure to raise the temperature. In the event that the instructions do not produce the desired result, checking the "no" square will bring up additional instructions provided by the expert. If the instructions correct the problem, the

---

PROCESS: COOKER, POTATO STARCH #6543

Select Problem

        Temperature   [X]

        Pressure   [ ]

        Flow   [ ]

        Color   [ ]

Figure 15-1. Problem selection.

```
┌─────────────────────────────────────────────────────┐
│  PROCESS: COOKER, POTATO STARCH #6543                 │
├─────────────────────────────────────────────────────┤
│  Temperature incorrect                                │
│                                                       │
│         Too High              ☐                      │
│                                                       │
│         Too Low               X                      │
│                                                       │
│         Wanders               ☐                      │
│                                                       │
│         Other                 ☐                      │
│                                                       │
└─────────────────────────────────────────────────────┘
```

Figure 15-2.   Problem definition.

```
┌─────────────────────────────────────────────────────┐
│  PROCESS: COOKER, POTATO STARCH #6543                 │
├─────────────────────────────────────────────────────┤
│                                                       │
│  Temperature Too Low                                  │
│                                                       │
│  1. Press overide button                              │
│  2. Tap "ADJUST HIGHER" bar twice                     │
│  3. Is temperature now correct?       yes ☐          │
│                                        no ☐          │
│  4. Explain what caused the low temperature ☐        │
│  5. HELP ......................☐                     │
│                                                       │
└─────────────────────────────────────────────────────┘
```

Figure 15-3.   Problem solution.

operator can ask for an explanation of the malfunction. In the event of an impending emergency, pressing "HELP" will bring up the crisis procedure window (and perhaps sound an alarm, or notify the supervisor). In actual practice, instructions might be expected to be more complicated than the example shown. A more typical instruction might read: "If the filters are becoming saturated, and the pump speed is near maximum, increase the airflow, add 2% additional filter aid, and monitor the pH carefully. If it exceeds 7.3, slowly close the feed valve."

A useful refinement to the above system would be the addition of a series of windows showing the overall process flow, accompanied by photographs of each piece of equipment. Additionally, process values and characteristics can be included. A disk containing all of these elements would be ideal for training new operators or product improvement teams.

By adding a data collection folder to the program, it would be a relatively uncomplicated step to include up-to-the minute control charts for each of the parameters (temperature, pressure, flow, color) for quality control and process improvement studies. Periodic printed control charts could be made available to the production supervisor and the quality control department.

COMPUTER-CONTROLLED PROCESSING

Before exploring the function of computers in quality control, a review of the principles of control might clarify this fairly complex subject. The three steps involved in quality control are measurement, evaluation, and response.

- The measurement can be made manually, visually, or by electrical, mechanical, biological, or chemical methods, or by combinations of these.
- The evaluation compares the measurement to a setpoint (the optimum value) and generates an error signal. In statistical quality control terms, this signal may be expressed as "within 3-sigma, take no action," or "beyond the 3-sigma control limits, take action." This class of evaluations may be accomplished manually or automatically. Another type of evaluation, usually conducted by automatic instruments, provides for one of three modes:
  1. *Proportional mode control:* Measures the magnitude of the error relative to an acceptable range.
  2. *Integral or reset mode control:* Measures the magnitude of the error and relates it to the interval of time over which the error occurred.
  3. *Derivative or rate mode of control:* Measures the rate at which the error is changing.
- The responses (or feedback signal) is generated by automatic instruments to reduce the error signal to zero. Older pneumatic and electronic controllers are classed as analog devices. Advances in digital computer technology have provided improved methods for control by the use of computer programs to evaluate the error. Commonly, the same modes of control are utilized: proportional, reset and rate.

A word of caution: Unless the statistical principles of process variation are considered when installing instrumented process control, overcontrol might be programmed into a computer system. This may produce a progression of frantic searching by the computer to reduce error in a process which may be perfectly satisfactory if left alone. In the worst case, this can produce defects both above and below the control limits. A few "seagoing" examples should illustrate overcontrol.

> To steer a canoe on a straight course, a correction is made with every stroke of the paddle. Occasionally, in a stiff breeze, an extra correction stroke might be required. Beyond this, an attempt to correct the oscillation of the canoe in mid-stroke would be overcontrol.

Once the course has been set and the rudder has been adjusted, a 100,000 ton tanker will continue in a straight line, requiring no correction for hours at a time.

A 10,000 ton freighter in a following sea (waves rolling in approximately the same direction of travel as the ship) will alternately swing slowly to the left and then to the right of the course if no rudder adjustment is made. Under this condition however, the overall direction of travel will be in a relatively straight line. If an attempt is made to correct the freighter's swing by steering in the opposite direction, the freighter will continue its swing for a few more minutes, and then will respond to both the rudder and the waves to overcorrect in the other direction. The net effect is to cause extremely wide oscillations, each larger than the preceding one. The overall direction of travel would be unpredictable.

Once a system is capable of operating successfully within satisfactory limits, using manual control or analog instrumentation, the next step many companies consider is the installation of computerized quality control. The ultimate goal is to reduce or eliminate the need for operator intervention. Continuous measurements of a process or product may result in large volumes of data which provide a guide to controlling quality. There are circumstances—perhaps a small company, or a short production run—where facilities and personnel may not be available to evaluate this data statistically. Here, too, the use of computer technology may simplify the control of quality.

Selection of the hardware and software to build a satisfactory control system can be accomplished in many ways. Perhaps the simplest method is to employ the services of a consultant who has successfully built effective computerized quality control systems for other companies in the food industry. Technical and production personnel should participate in the development and receive training in the program as it is created. If personnel in quality control and engineering are available, the company may elect to provide them with the specialized training to choose their own program from the hundreds available. The advantage of having an expert or two on the payroll is obvious; conversely, the disadvantage is the possible promotion or loss of these employees. The use of a team to assist the technical personnel may slow the process somewhat but should provide the cooperation necessary to make any new system workable.

To construct a new or improved computer-controlled process, a number of lengthy observations and calculations must be prepared. Each stage of the process has both inputs and outputs from and to many associated process stages. These must be measured and arranged in tabular (or graphical) form so that the optimum level of each output can be calculated. Where there are many interrelated operations to the overall process, this exploratory step can be very lengthy. The next procedure used by many programmers is to connect a series of control instruments which simulate the process steps above and which can be operated at various settings to validate the calculations.

For years, these two initial steps had to be performed manually, and usually at great expense. Software tools are now available to perform the analysis and simulations more quickly. These programs can then be combined to demonstrate how the operation would eventually be computer controlled, and how realistically the theoretical calculations were prepared. By running a series of simulations, these programs can accurately forecast the expected output of the process under a number of varying inputs.

Installing a computerized system of production or quality control requires major decisions in three areas:

1. What hardware will do the job?
2. What software will make it work?
3. How do we prepare the work force?

Successful installations are rarely those which attempt to cover every aspect of the company's operations with a single blanket. Once top management is convinced that computer control is the way to go, it is essential that a management team be created with the understanding that they will start small, growing as the payback is demonstrated. The team must also be prepared to solve many new production and quality problems, because the ease and speed of computerized data collection will most surely disclose several problems not previously recognized. Unless these are handled as they occur, the new system will probably fail. If problems are not solved promptly, the same "red lights" will continue to flash every day, and the effectiveness of the computer will be lost.

Some of the obvious advantages to computer-guided operations are their assistance in eliminating operator variability, avoiding use of the wrong or obsolete materials, instructing operators on performance of their jobs, and entering data directly from the measuring device output (calipers, pH meters, thermometers, balances, HPLC). For some applications, data can be color-coded as it is entered in order to inform the operator if the process is satisfactory (green), marginal (yellow), or in trouble (red). For simplicity, the data can be entered on a speedometer dial, a thermometer scale, or stair-step diagram to assist in interpreting the measurement before the numbers are entered into the statistical data bank for further processing and recording. For additional value, the data calculations may be shown on the computer screen as control charts in real time to assist the operator, the supervisor, and the manager in evaluating the process while it is underway. Additionally, the program can continuously produce an updated histogram for daily, monthly, or other comparisons. Of equal importance, a Pareto analysis of the problems encountered in the process can be simultaneously constructed and updated. This would permit anyone interested in the process to determine the most critical problems, at any time. Inventory control may continue to be a function of the warehouse, but the production floor computers may also be programmed to control inventory as production progresses. Where maintenance scheduling is critical, the process computer may also be used to keep track of machine-hours, flashing a notice when maintenance performance is approaching. Even in the control laboratory, a dozen different quality tests can be available for display on the computer screen simultaneously, along with control or specification limits, so that out-of-limits data can be immediately signaled for a collection of analyses such as color, moisture, pH, flow, solubility, texture, alkalinity, and so on. If desired, all of the information from warehouse, production floor, quality control laboratory, shipping, and accounting can be "piped" to the CEO's office for real-time evaluation of each department as well as the entire operation. Finally, each analytical tool performed by the computer may be programmed to flash a warning light or screen message when an undesirable condition occurs.

Data Input

In addition to the necessity of providing guidance in operating a process, there are many other reasons for taking data describing a product's quality characteristics. Some of the common ones follow:

1. A customer may require data as part of the purchase order. The requirement may be contractual to certify a lot's conformance to specifications, or it may be considered important to the customer's manufacturing or marketing systems. The data may or may not have any realistic value to the customer.
2. The CEO or the legal department may wish to have ample files as backup in the event of a product quality failure, resulting in a recall or a lawsuit.
3. A data bank of purchased raw material performance may be a useful tool in negotiating price.
4. A confused or untrained management might ask for volumes of assorted data with the hope of finding a clue to areas for quality improvement.
5. Defect data might be accumulated by incoming shipments to assist in selection of a supplier.
6. A large file on competitive product quality might be a useful tool for benchmarking purposes.

It is likely that there are another half-dozen reasons for collecting data, but there are four general levels of information which seem to be commonly required by all companies:

- Top management wants information on an $8\frac{1}{2} \times 11$ piece of paper which tells them "what's going on," an overview of what is happening.
- On the shop floor, the workers need the type of information which will give them data regarding the variability of the process, allowing them to take action in real time.
- Middle management (QC or Engineering) uses parcels of information to statistically analyze and compare—after the fact. They may be interested in comparing shifts, machines, products, processes, production days, operators, and material supplier performance.
- Product and process improvement teams or Research and Development usually require large masses of past data in their efforts to reach their goals.

Middle management needs considerably more information than is required for immediate use by workers on the production floor. The data may have to be tagged with shift, machine, product, serial number, lot size labels for further analysis.

The major advantages of electronic data collection and transmission are accuracy and speed. Collecting data by individual hand measurement, writing the data onto a form, calculating or summarizing, and copying it onto a chart presents opportunities for making an error (is that figure a 7 or a 5?) and requires far more time to perform than does an electronic system. It has been found in a detailed study that mistakes in writing down and reading numbers can run as high as 12%. A major improvement over copying data by hand would be to enter the information through a computer keyboard, although the opportunity for errors still exists. Variable data can be collected from virtually any electronic output—gages, calipers, pH meters, chromatographs, scales, and so on—and be transmitted error-free through one of many relatively inexpensive "black boxes" to computers for display, storage, or further processing.

One effective method of collecting attribute data is the use of a bar code. Hand-held devices are available which can be programmed to read bar codes, store, calculate, or transmit the information to a master computer. They can be programmed to perform specific tasks, or they can be purchased preprogrammed for general use. In addition to the obvious functions of a bar code for identifying raw materials on the receiving dock and finished product in the warehouse, they are also used to classify blocks of data collected at specific stages on the production line.

Before investing in an electronic data collection system, one must be sure that there is no proprietary function in the black box which will make it impossible to use for a particular operation. There is a myriad of products available, but some of them are programmed to handle a specific function which might not be applicable to the process under consideration.

Some hand-held computers, such as personal organizer units, are available which can be programmed to contain the day's schedule, which accept pen input for notes or graphics, and which can be down-loaded to desktop computers for further processing or recording.

Modules are available which can accept voice data—in any language. Others can speak to the operator in phrases or whole pages which are highlighted on the screen. This capability is of value where the employees are not proficient in reading, or need translation.

Through the use of video cameras and video recorders, tapes of a process can be readily prepared while in operation, to be used as a training film for other employees at the same or different locations. If required, instantaneous transfer of such tapes can be made using a modem and telephone lines, thus permitting interchange of ideas between widely separated plants while operating or modifying a process.

Sensing Devices

A sensor is a device which reacts to a physical or chemical stimulus to produce a corresponding electrical signal proportional to the physical parameters measured.

As generally understood, a sensor is another word for a transducer. In the preceding example of an Expert System, four such sensors were used to control a process for cooking potato starch: temperature detector, pressure measuring device, flow rate instrument, and color analyzer. See Figure 15-4. Each of these sensors responded to specific stimuli, and converted the relative response to an electrical signal which, in turn, was sent ultimately to the computer at the workstation. The signal was then processed by a computer program and returned to an operating device (such as a solenoid valve) which would appropriately adjust the physical condition as required to maintain control. In the event of a physical condition beyond the control limits for which the program was defined, an out-of-limits notice was issued on screen for the operator's attention (the temperature, in the example shown). As an additional tool for the operator, an overall pictorial of the process under his control may be included in the program as illustrated below. It shows the major equipment and the relative location of the sensors and controllers.

Sensors are available for countless conditions and respond to stimuli using a wide variety of principles. Temperature sensors, for example, may be classified as thermocouples (TC), resistance temperature detectors (RTD), thermistors, and IC sensors. Other sensors are listed in Table 15-2.

Flow meters can be based on differential pressure, magnetic sensor, vortex, coriolis effect, thermal mass, positive displacement, turbine, ultrasonic, and other sensors. Level sensors may also be triggered by capacitance/radio frequency, hydrostatic pressure, float switch, photocell, radar/microwave, ultrasonics, load cell, conductance instrument, and point switches.

Automatic Control

There are a number of precautions which should precede the discussion of automatic control of a process. First, equipment may unexpectedly fail to perform for a variety of reasons, some of which might not be at all apparent. Valves which have not been moved for long periods have been known to jam in their last position. Light sensing devices, such as colorimeters or refractometers may develop

Figure 15-4.   Process flow controls.

Table 15-2.   Sensors

Sensors	Measurement
Infrared, conductivity	Acid
Polymer resins	Aroma
Bio-reactor	BOD (high speed)
Turbidimeters	Clarity
Colorimeter	Color
Infrared	Composition (chemical)
Electrical resistance	Concentration
X-ray	Contamination, solid
Gamma ray, scales	Count
Gamma rays	Density
Reflectometer	Defect (visual)
Infrared	Fat
Gamma ray	Fill level
Strain gage	Force
Differential pressure	Flow rate
Polarized or IR light	Glass chips
Strain gage/Brinell	Hardness
Visible light	Labels (missing, faulty)
Float, strain gage	Level
Photocell	Light
Proximity switch	Motion
Strain gage	Mass (weight)
Infrared	Moisture
pH Cell	pH
Pressure transducer	Pressure
Photocell	Seal area contamination
Microphone	Sound
Viscometer	Shear
Refractometer	Solids
Infrared	Sugar
Infrared	Sulfur dioxide
Thermocouple	Temperature
Conductivity	Total acid
Rheometer	Viscosity

cracked, etched, scaled, or clouded lenses from product or environmental contact. Growth of plant life or insect colonies can clog flow measuring devices. Electrical power fluctuations may produce erroneous signals. Strain gages or spring-loaded devices may require recalibration due to age or wear. Vibration or temperature has been known to warp indicating arrows on meters. Barometric pressure changes may cause false reading in vacuum-activated measuring devices. Humidity can affect electrical circuits.

This is not to say that automatic controls cannot be trusted, but it does suggest that those responsible for quality control should be aware of the requirement that the ultimate success or failure of an automatic system depends on routinely maintained and calibrated equipment. In the final analysis, the success of a system is measured by the consistent, predictable uniform quality of the finished product.

Figure 15-5 contains the major elements of a computer-based process controlled system:

Sensor: Senses a physical or chemical condition, converts this impulse to an electrical signal proportional to the condition measured.

Signal Conditioner: Converts the sensor signal into a form acceptable to the Data Analyzer. In some instances, the signal may be amplified or filtered. The amplification of the signal received should be such that its maximum voltage range equals the maximum input of the Data Analyzer. In some cases, it may be required to reduce the electrical signal to protect the computer. Filtering may be desirable to remove unwanted signals from certain sensors. Some signals are logarithmic in nature, and their output should be reduced to a linear voltage for maximum sensitivity.

Data Acquisition Board (DAB): Accepts signals from a number of channels, at specified sampling rates and types of inputs. The sampling rates are often determined by the ability of a DAB to multiplex, that is, switch from channel to channel. These boards are available in a wide array of capabilities and outputs.

Computer: Most computers have sufficient computational power to handle the output of available DABs. If necessary, however, special analysis hardware is available. Conclusions can be performed in the millions per second, allowing a computer to handle many inputs from the DAB at extremely high speed.

Software: Some companies might prefer to develop their own software to program the hardware. It is probably quicker (and certainly simpler) to purchase driver software from the manufacturer of the DAB. The use of application software will add presentation and analysis capabilities to the driver software.

Activator: Outputs from the computer in automatic control systems are primarily concerned with operating control devices such as valves, motor

Figure 15-5. Computer-based process control system.

drives, positioning arms, mixers, fans and other powered equipment. Expert System inputs/outputs can also be programmed into some systems. As discussed above, the computer quality control calculating output can take the form of data compilations, control charts, run charts, and many other graphs.

As mentioned above, the computer makes data analysis both easy and rapid. In addition to the control charts which can be instantly calculated and viewed by the line operator, the same data can be down-loaded or connected through computer networks to quality control and process engineers for further analysis.

Consider the simplified example in Figure 15-6. A small company manufactures a family of drink mixes, using a simple system as shown in the flow chart. The raw ingredients are weighed manually and dumped into a screw blender which discharges through a feed bin into a four-scale filler head. The products are packaged in composite foil canisters and cased after labeling. Net weights are checked occasionally. The major reason for check-weighing is to reduce product giveaway and to avoid the possibility of creating seam cutovers from overfilling.

The principle customer has specified that the net weights should average 454 grams, with an acceptable range of 445–463 grams. Over the years, this customer has complained about underweights on several occasions and has returned the offending shipments for reworking—that is, hand sorting and removing the underweight containers. Eventually, this customer tired of the delays resulting from the weight problem and threatened to take his business elsewhere. The customer suggested contacting the Smathers Scale Company, which manufactured automatic checkweighers.

The Smathers representative brought in an automatic checkweigher and data computer for a demonstration. While it was being assembled and wired into the line, the representative took 50 consecutive samples of orange drink mix off the

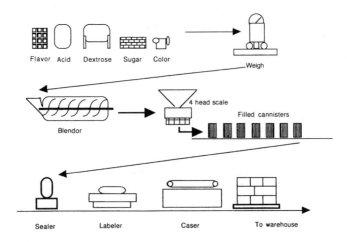

Figure 15-6. Drink mix flow chart.

line as produced. Over the next several hours he weighed each can, subtracting the tare weight, then calculated the process capability based on these samples, using a hand-held calculator.

The calculations showed that over an extended run, the process would be expected to produce nearly 10% below the lower specification limit of 445 grams, and about 6% above the upper specification limit of 463 grams. This came as a shock to the owner, who had assumed that the four filler scales were always reliable.

The following day, the automatic Smathers equipment was ready to demonstrate. The checkweigher system gross weighed each canister as it passed over a strain gauge, adjusting for the tare. The electrical signal generated was picked up by a computer, which stored the data, and was programmed to perform a number of statistical functions. The line speed was running at the usual 45 canisters per minute, and in less than 2 minutes, nearly 100 weighings were recorded in the computer. Smathers then pressed the PROCESS CAPABILITY button on the console, and selected the CALCULATE option. A number of statistical calculations appeared on the computer screen within seconds:

> Sample size 50
> Mean 453.206
> Standard deviation. 6.32257
> 3 Sigma limits . . . . 434.238 to 472.174
> Lower specification limit. 445
> Upper specification limit. 463
> CPK. 0.43263
> CP Index 0.474491
> % Under LSL of 445 . . . . 9.7%
> % Over USL of 463. . . . . 6.1%

To further demonstrate the effectiveness of the system, Smathers then selected the GRAPH option, and a graph appeared on the computer screen within a few seconds (Fig. 15-7). The data, calculations, and graph all showed that the system was not capable of satisfying the customer's requirements.

The graph and calculations convinced the owner that the scales were in need of maintenance. The counterweights were adjusted to the precise delivery weights, and the knife-edge scale bearings were honed. With the assistance of the Smathers representative, another 50 samples were automatically weighed and recorded in the computer. The resulting chart (Fig. 15-8) showed that the line was now capable of producing at least the minimum weight, although some overweights could still be expected. Changing the line speed would probably permit some control over the average weight value, but the process was marginal.

After much discussion, it was decided that better density control could narrow the weight range, and combined with the improved line maintenance, perhaps good control might be attained. The density could be adjusted by varying the blender time for each batch. Once the ideal blender duration was established, an-

Figure 15-7.   Initial capability analysis.

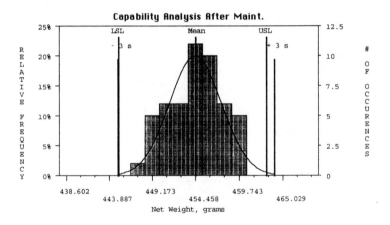

Figure 15-8.   Capability analysis after maintenance.

other set of data was fed into the computer and analyzed. The results are pre-sented in Figure 15-9.

The PROCESS CAPABILITY–CALCULATE buttons were activated, produc-ing the following data:

    Sample size 50
    Mean 453.624
    Standard deviation 2.65196
    3 Sigma limits . . . . 445.668 to 461.58
    Lower specification limit 445
    Upper specification limit 463
    CPK 1.08398
    CP Index 1.13124

Figure 15-9. Capability analysis with density control.

Figure 15-10. Net weight control chart (with maintenance, and density control).

% Under LSL of 445. . . . . 0.1%
% Over USL of 463 . . . . 0.0%

To demonstrate the flexibility of the Smathers computer programs (and to further assist in closing the sale), some of the other features were demonstrated by the simple pressing of the appropriate computer panel buttons. Using the data from the capability study with density control (the final test conditions), the control charts presented in Figures 15-10 and 15-11 were instantly constructed:

By depressing the DATA button, the 50 data points used to construct the preceding control charts were printed out in seconds. Two other charts were also demonstrated: trend analysis and *X/Y* chart.

This simple example has demonstrated the speed with which computers can solve production problems, calculate and graph results of changes, and provide high-speed calculations in process improvement studies.

Figure 15-11.   Net weight range control chart.

SUMMARY

To summarize, a computer system can be installed on the factory floor to permit the operator to enter and observe data with reference to standards. The data may be combined with previous information to produce control charts for review by the operator, the supervisor, or the manager to evaluate the process at any time with respect to current or prior performance. The computer might serve as an expert with push-button information to guide the operator in performance of his work. Various signals or alarms may be included to notify the operator of the need for remedial action. The signals received by the computer may bypass the operator and automatically trigger control reactions to operate the process at optimum conditions. Finally, the data may be stored or transferred to other computer banks for further analysis or for historical purposes.

Appendix

Table A-1.   Summation of Terms of the Poisson Exponential Binomial Distribution Limit

Entries in the table, when multiplied by 10^{-3}, represent the probabilities of c or fewer defects, defectives, or occurrences when the expected or average number is np'.

c' or np'	0	1	2	3	4	5	6	7	8	9
0.02	980	1,000								
0.04	961	999	1,000							
0.06	942	998	1,000							
0.08	923	997	1,000							
0.10	905	995	1,000							
0.15	861	990	999	1,000						
0.20	819	982	999	1,000						
0.25	779	974	998	1,000						
0.30	741	963	996	1,000						
0.35	705	951	994	1,000						
0.40	670	938	992	999	1,000					
0.45	638	925	989	999	1,000					
0.50	607	910	986	998	1,000					
0.55	577	894	982	998	1,000					
0.60	549	878	977	997	1,000					
0.65	522	861	972	996	999	1,000				
0.70	497	844	966	994	999	1,000				
0.75	472	827	959	993	999	1,000				
0.80	449	809	953	991	999	1,000				
0.85	427	791	945	989	998	1,000				
0.90	407	772	937	987	998	1,000				
0.95	387	754	929	984	997	1,000				
1.00	368	736	920	981	996	999	1,000			
1.1	333	699	900	974	995	999	1,000			
1.2	301	663	879	966	992	998	1,000			
1.3	273	627	857	957	989	998	1,000			
1.4	247	592	833	946	986	997	999	1,000		
1.5	223	558	809	934	981	996	999	1,000		
1.6	202	525	783	921	976	994	999	1,000		
1.7	183	493	757	907	970	992	998	1,000		
1.8	165	463	731	891	964	990	997	999	1,000	
1.9	150	434	704	875	956	987	997	999	1,000	
2.0	135	406	677	857	947	983	995	999	1,000	

Credit: Eugene L. Grant, *Statistical Quality Control*, 3rd ed. (New York: McGraw-Hill Book Company, 1964).

Table A-1. *(continued)*

c' or np' \ c	0	1	2	3	4	5	6	7	8	9
2.2	111	355	623	819	928	975	993	998	1,000	
2.4	091	308	570	779	904	964	988	997	999	1,000
2.6	074	267	518	736	877	951	983	995	999	1,000
2.8	061	231	469	692	848	935	976	992	998	999
3.0	050	199	423	647	815	916	966	988	996	999
3.2	041	171	380	603	781	895	955	983	994	998
3.4	033	147	340	558	744	871	942	977	992	997
3.6	027	126	303	515	706	844	927	969	988	996
3.8	022	107	269	473	668	816	909	960	984	994
4.0	018	092	238	433	629	785	889	949	979	992
4.2	015	078	210	395	590	753	867	936	972	989
4.4	012	066	185	359	551	720	844	921	964	985
4.6	010	056	163	326	513	686	818	905	955	980
4.8	008	048	143	294	476	651	791	887	944	975
5.0	007	040	125	265	440	616	762	867	932	968
5.2	006	034	109	238	406	581	732	845	918	960
5.4	005	029	095	213	373	546	702	822	903	951
5.6	004	024	082	191	342	512	670	797	886	941
5.8	003	021	072	170	313	478	638	771	867	929
6.0	002	017	062	151	285	446	606	744	847	916

c' or np' \ c	10	11	12	13	14	15	16
2.8	1,000						
3.0	1,000						
3.2	1,000						
3.4	999	1,000					
3.6	999	1,000					
3.8	998	999	1,000				
4.0	997	999	1,000				
4.2	996	999	1,000				
4.4	994	998	999	1,000			
4.6	992	997	999	1,000			
4.8	990	996	999	1,000			
5.0	986	995	998	999	1,000		
5.2	982	993	997	999	1,000		
5.4	977	990	996	999	1,000		
5.6	972	988	995	998	999	1,000	
5.8	965	984	993	997	999	1,000	
6.0	957	980	991	996	999	999	1,000

(continued)

Table A-1. *(continued)*

c ć or np′	0	1	2	3	4	5	6	7	8	9
6.2	002	015	054	134	259	414	574	716	826	902
6.4	002	012	046	119	235	384	542	687	803	886
6.6	001	010	040	105	213	355	511	658	780	869
6.8	001	009	034	093	192	327	480	628	755	850
7.0	001	007	030	082	173	301	450	599	729	830
7.2	001	006	025	072	156	276	420	569	703	810
7.4	001	005	022	063	140	253	392	539	676	788
7.6	001	004	019	055	125	231	365	510	648	765
7.8	000	004	016	048	112	210	338	481	620	741
8.0	000	003	014	042	100	191	313	453	593	717
8.5	000	002	009	030	074	150	256	386	523	653
9.0	000	001	006	021	055	116	207	324	456	587
9.5	000	001	004	015	040	089	165	269	392	522
10.0	000	000	003	010	029	067	130	220	333	458

	10	11	12	13	14	15	16	17	18	19
6.2	949	975	989	995	998	999	1,000			
6.4	939	969	986	994	997	999	1,000			
6.6	927	963	982	992	997	999	999	1,000		
6.8	915	955	978	990	996	998	999	1,000		
7.0	901	947	973	987	994	998	999	1,000		
7.2	887	937	967	984	993	997	999	999	1,000	
7.4	871	926	961	980	991	996	998	999	1,000	
7.6	854	915	954	976	989	995	998	999	1,000	
7.8	835	902	945	971	986	993	997	999	1,000	
8.0	816	888	936	966	983	992	996	998	999	1,000
8.5	763	849	909	949	973	986	993	997	999	999
9.0	706	803	876	926	959	978	989	995	998	999
9.5	645	752	836	898	940	967	982	991	996	998
10.0	583	697	792	864	917	951	973	986	993	997

	20	21	22
8.5	1,000		
9.0	1,000		
9.5	999	1,000	
10.0	998	999	1,000

Table A-1. *(continued)*

c c' or np'	0	1	2	3	4	5	6	7	8	9
10.5	000	000	002	007	021	050	102	179	279	397
11.0	000	000	001	005	015	038	079	143	232	341
11.5	000	000	001	003	011	028	060	114	191	289
12.0	000	000	001	002	008	020	046	090	155	242
12.5	000	000	000	002	005	015	035	070	125	201
13.0	000	000	000	001	004	011	026	054	100	166
13.5	000	000	000	001	003	008	019	041	079	135
14.0	000	000	000	000	002	006	014	032	062	109
14.5	000	000	000	000	001	004	010	024	048	088
15.0	000	000	000	000	001	003	008	018	037	070

	10	11	12	13	14	15	16	17	18	19
10.5	521	639	742	825	888	932	960	978	988	994
11.0	460	579	689	781	854	907	944	968	982	991
11.5	402	520	633	733	815	878	924	954	974	986
12.0	347	462	576	682	772	844	899	937	963	979
12.5	297	406	519	628	725	806	869	916	948	969
13.0	252	353	463	573	675	764	835	890	930	957
13.5	211	304	409	518	623	718	798	861	908	942
14.0	176	260	358	464	570	669	756	827	883	923
14.5	145	220	311	413	518	619	711	790	853	901
15.0	118	185	268	363	466	568	664	749	819	875

	20	21	22	23	24	25	26	27	28	29
10.5	997	999	999	1,000						
11.0	995	998	999	1,000						
11.5	992	996	998	999	1,000					
12.0	988	994	997	999	999	1,000				
12.5	983	991	995	998	999	999	1,000			
13.0	975	986	992	996	998	999	1,000			
13.5	965	980	989	994	997	998	999	1,000		
14.0	952	971	983	991	995	997	999	999	1,000	
14.5	936	960	976	986	992	996	998	999	999	1,000
15.0	917	947	967	981	989	994	997	998	999	1,000

(continued)

Table A-1. *(continued)*

c' or np' \ c	4	5	6	7	8	9	10	11	12	13
16	000	001	004	010	022	043	077	127	193	275
17	000	001	002	005	013	026	049	085	135	201
18	000	000	001	003	007	015	030	055	092	143
19	000	000	001	002	004	009	018	035	061	098
20	000	000	000	001	002	005	011	021	039	066
21	000	000	000	000	001	003	006	013	025	043
22	000	000	000	000	001	002	004	008	015	028
23	000	000	000	000	000	001	002	004	009	017
24	000	000	000	000	000	000	001	003	005	011
25	000	000	000	000	000	000	001	001	003	006

	14	15	16	17	18	19	20	21	22	23
16	368	467	566	659	742	812	868	911	942	963
17	281	371	468	564	655	736	805	861	905	937
18	208	287	375	469	562	651	731	799	855	899
19	150	215	292	378	469	561	647	725	793	849
20	105	157	221	297	381	470	559	644	721	787
21	072	111	163	227	302	384	471	558	640	716
22	048	077	117	169	232	306	387	472	556	637
23	031	052	082	123	175	238	310	389	472	555
24	020	034	056	087	128	180	243	314	392	473
25	012	022	038	060	092	134	185	247	318	394

	24	25	26	27	28	29	30	31	32	33
16	978	987	993	996	998	999	999	1,000		
17	959	975	985	991	995	997	999	999	1,000	
18	932	955	972	983	990	994	997	998	999	1,000
19	893	927	951	969	980	988	993	996	998	990
20	843	888	922	948	966	978	987	992	995	997
21	782	838	883	917	944	963	976	985	991	994
22	712	777	832	877	913	940	959	973	983	989
23	635	708	772	827	873	908	936	956	971	981
24	554	632	704	768	823	868	904	932	953	960
25	473	553	629	700	763	818	863	900	929	950

	34	35	36	37	38	39	40	41	42	43
19	999	1,000								
20	999	999	1,000							
21	997	998	999	999	1,000					
22	994	996	998	999	999	1,000				
23	988	993	996	997	999	999	1,000			
24	979	987	992	995	997	998	999	999	1,000	
25	966	978	985	991	994	997	998	999	999	1,000

Table A-2. Areas under the Normal Probability Curve

$$Z_0 = \frac{x - \mu}{\sigma}$$

Area to the right of z (or to the left of $-z$), or the probability of a random value of z exceeding the marginal value

z_0	.00	.01	.02	.03	.04	.05	.06	.07	.08	.09
0.0	.500	.496	.492	.488	.484	.480	.476	.472	.468	.464
0.1	.460	.456	.452	.448	.444	.440	.436	.433	.429	.425
0.2	.421	.417	.413	.409	.405	.401	.397	.394	.390	.386
0.3	.382	.378	.374	.371	.367	.363	.359	.356	.352	.348
0.4	.345	.341	.337	.334	.330	.326	.323	.319	.316	.312
0.5	.309	.305	.302	.298	.295	.291	.288	.284	.281	.278
0.6	.274	.271	.268	.264	.261	.258	.255	.251	.248	.245
0.7	.242	.239	.236	.233	.230	.227	.224	.221	.218	.215
0.8	.212	.209	.206	.203	.200	.198	.195	.192	.189	.187
0.9	.184	.181	.179	.176	.174	.171	.169	.166	.164	.161
1.0	.159	.156	.154	.152	.149	.147	.145	.142	.140	.138
1.1	.136	.133	.131	.129	.127	.125	.123	.121	.119	.117
1.2	.115	.113	.111	.109	.107	.106	.104	.102	.100	.099
1.3	.097	.095	.093	.092	.090	.089	.087	.085	.084	.082
1.4	.081	.079	.078	.076	.075	.074	.072	.071	.069	.068
1.5	.067	.066	.064	.063	.062	.061	.059	.058	.057	.056
1.6	.055	.054	.053	.052	.051	.049	.048	.047	.046	.046
1.7	.045	.044	.043	.042	.041	.040	.039	.038	.038	.037
1.8	.036	.035	.034	.034	.033	.032	.031	.031	.030	.029
1.9	.029	.028	.027	.027	.026	.026	.025	.024	.024	.023
2.0	.023	.022	.022	.021	.021	.020	.020	.019	.019	.018
2.1	.018	.017	.017	.017	.016	.016	.015	.015	.015	.014
2.2	.014	.014	.013	.013	.013	.012	.012	.012	.011	.011
2.3	.011	.010	.010	.010	.010	.009	.009	.009	.009	.008
2.4	.008	.008	.008	.008	.007	.007	.007	.007	.007	.006
2.5	.006	.006	.006	.006	.006	.005	.005	.005	.005	.005
2.6	.005	.005	.004	.004	.004	.004	.004	.004	.004	.004
2.7	.003	.003	.003	.003	.003	.003	.003	.003	.003	.003
2.8	.003	.002	.002	.002	.002	.002	.002	.002	.002	.002
2.9	.002	.002	.002	.002	.002	.002	.002	.001	.001	.001

z_0	DETAIL OF TAIL ($._2$135, FOR EXAMPLE, MEANS .00135)									
2.	$._1$228	$._1$179	$._1$139	$._1$107	$._2$820	$._2$621	$._2$466	$._2$347	$._2$256	$._2$187
3.	$._2$135	$._3$968	$._3$687	$._3$483	$._3$337	$._3$233	$._3$159	$._3$108	$._4$723	$._4$481
4.	$._4$317	$._4$207	$._4$133	$._5$854	$._5$541	$._5$340	$._5$211	$._5$130	$._6$793	$._6$479
5.	$._6$287	$._6$170	$._7$996	$._7$579	$._7$333	$._7$190	$._7$107	$._8$599	$._8$332	$._8$182
	0.0	0.1	0.2	0.3	0.4	0.5	0.6	0.7	0.8	0.9

Credit: Ronald J. Wonnacott and Thomas H. Wonnacott, *Introductory Statistics*, 4th ed. (New York: John Wiley & Sons, 1985).

Table A-3.   Values of *t*

v \ *a*	\multicolumn{7}{c}{Two-tail Critical Values}
	0.50	0.25	0.10	0.05	0.025	0.01	0.005
1	1.00000	2.4142	6.3138	12.706	25.452	63.657	127.32
2	0.81650	1.6036	2.9200	4.3027	6.2053	9.9248	14.089
3	0.76489	1.4226	2.3534	3.1825	4.1765	5.8409	7.4533
4	0.74070	1.3444	2.1318	2.7764	3.4954	4.6041	5.5976
5	0.72669	1.3009	2.0150	2.5706	3.1634	4.0321	4.7733
6	0.71756	1.2733	1.9432	2.4469	2.9687	3.7074	4.3168
7	0.71114	1.2543	1.8946	2.3646	2.8412	3.4995	4.0293
8	0.70639	1.2403	1.8595	2.3060	2.7515	3.3554	3.8325
9	0.70272	1.2297	1.8331	2.2622	2.6850	3.2498	3.6897
10	0.69981	1.2213	1.8125	2.2281	2.6338	3.1693	3.5814
11	0.69745	1.2145	1.7959	2.2010	2.5931	3.1058	3.4966
12	0.69548	1.2089	1.7823	2.1788	2.5600	3.0545	3.4284
13	0.69384	1.2041	1.7709	2.1604	2.5326	3.0123	3.3725
14	0.69242	1.2001	1.7613	2.1448	2.5096	2.9768	3.3257
15	0.69120	1.1967	1.7530	2.1315	2.4899	2.9467	3.2860
16	0.69013	1.1937	1.7459	2.1199	2.4729	2.9208	3.2520
17	0.68919	1.1910	1.7396	2.1098	2.4581	2.8982	3.2225
18	0.68837	1.1887	1.7341	2.1009	2.4450	2.8784	3.1966
19	0.68763	1.1866	1.7291	2.0930	2.4334	2.8609	3.1737
20	0.68696	1.1848	1.7247	2.0860	2.4231	2.8453	3.1534
21	0.68635	1.1831	1.7207	2.0796	2.4138	2.8314	3.1352
22	0.68580	1.1816	1.7171	2.0739	2.4055	2.8188	3.1188
23	0.68531	1.1802	1.7139	2.0687	2.3979	2.8073	3.1040
24	0.68485	1.1789	1.7109	2.0639	2.3910	2.7969	3.0905
25	0.68443	1.1777	1.7081	2.0595	2.3846	2.7874	3.0782
26	0.68405	1.1766	1.7056	2.0555	2.3788	2.7787	3.0669
27	0.68370	1.1757	1.7033	2.0518	2.3734	2.7707	3.0565
28	0.68335	1.1748	1.7011	2.0484	2.3685	2.7633	3.0469
29	0.68304	1.1739	1.6991	2.0452	2.3638	2.7564	3.0380
30	0.68276	1.1731	1.6973	2.0423	2.3596	2.7500	3.0298
40	0.68066	1.1673	1.6839	2.0211	2.3289	2.7045	2.9712
60	0.67862	1.1616	1.6707	2.0003	2.2991	2.6603	2.9146
120	0.67656	1.1559	1.6577	1.9799	2.2699	2.6174	2.8599
∞	0.67449	1.1503	1.6449	1.9600	2.2414	2.5758	2.8070

v	0.25	0.125	0.05	0.025	0.0125	0.005	0.0025

a — \multicolumn{7}{c}{One-tail Critical Values}

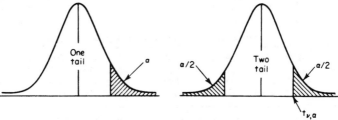

Credit: E. S. Pearson, "Critical Values of Student's *t* Distribution," *Biometrika* 32 (1941): 168–181.

Table A-4. Binomial Coefficients

n	$\binom{n}{0}$	$\binom{n}{1}$	$\binom{n}{2}$	$\binom{n}{3}$	$\binom{n}{4}$	$\binom{n}{5}$	$\binom{n}{6}$	$\binom{n}{7}$	$\binom{n}{8}$	$\binom{n}{9}$	$\binom{n}{10}$
0	1										
1	1	1									
2	1	2	1								
3	1	3	3	1							
4	1	4	6	4	1						
5	1	5	10	10	5	1					
6	1	6	15	20	15	6	1				
7	1	7	21	35	35	21	7	1			
8	1	8	28	56	70	56	28	8	1		
9	1	9	36	84	126	126	84	36	9	1	
10	1	10	45	120	210	252	210	120	45	10	1
11	1	11	55	165	330	462	462	330	165	55	11
12	1	12	66	220	495	792	924	792	495	220	66
13	1	13	78	286	715	1287	1716	1716	1287	715	286
14	1	14	91	364	1001	2002	3003	3432	3003	2002	1001
15	1	15	105	455	1365	3003	5005	6435	6435	5005	3003
16	1	16	120	560	1820	4368	8008	11440	12870	11440	8008
17	1	17	136	680	2380	6188	12376	19448	24310	24310	19448
18	1	18	153	816	3060	8568	18564	31824	43758	48620	43758
19	1	19	171	969	3876	11628	27132	50388	75582	92378	92378
20	1	20	190	1140	4845	15504	38760	77520	125970	167960	184756

$$C_r^n = \binom{n}{r} = \frac{n!}{r!(n-r)!} = \binom{n}{n-r}, \quad \binom{n}{0} = 1.$$

$$(q+p)^n = q^n + C_1^n pq^{n-1} + \cdots + C_r^n p^r q^{n-r} + \cdots + p^n.$$

Example: $(q+p)^4 = q^4 + 4q^3p + 6q^2p^2 + 4qp^3 + p^4$

Table A-5. Random Numbers

10 27 53 96 23	71 50 54 36 23	54 31 04 82 98	04 14 12 15 09	26 78 25 47 47
28 41 50 61 88	64 85 27 20 18	83 36 36 05 56	39 71 65 09 62	94 76 62 11 89
34 21 42 57 02	59 19 18 97 48	80 30 03 30 98	05 24 67 70 07	84 97 50 87 46
61 81 77 23 23	82 82 11 54 08	53 28 70 58 96	44 07 39 55 43	42 34 43 39 28
61 15 18 13 54	16 86 20 26 88	90 74 80 55 09	14 53 90 51 17	52 01 63 01 59
91 76 21 64 64	44 91 13 32 97	75 31 62 66 54	84 80 32 75 77	56 08 25 70 29
00 97 79 08 06	37 30 28 59 85	53 56 68 53 40	01 74 39 59 73	30 19 99 85 48
36 46 18 34 94	75 20 80 27 77	78 91 69 16 00	08 43 18 73 68	67 69 61 34 25
88 98 99 60 50	65 95 79 42 94	93 62 40 89 96	43 56 47 71 66	46 76 29 67 02
04 37 59 87 21	05 02 03 24 17	47 97 81 56 51	92 34 86 01 82	55 51 33 12 91
63 62 06 34 41	94 21 78 55 09	72 76 45 16 94	29 95 81 83 83	79 88 01 97 30
78 47 23 53 90	34 41 92 45 71	09 23 70 70 07	12 38 92 79 43	14 85 11 47 23
87 68 62 15 43	53 14 36 59 25	54 47 33 70 15	59 24 48 40 35	50 03 42 99 36
47 60 92 10 77	88 59 53 11 52	66 25 69 07 04	48 68 64 71 06	61 65 70 22 12
56 88 87 59 41	65 28 04 67 53	95 79 88 37 31	50 41 06 94 76	81 83 17 16 33
02 57 45 86 67	73 43 07 34 48	44 26 87 93 29	77 09 61 67 84	06 69 44 77 75
31 54 14 13 17	48 62 11 90 60	68 12 93 64 28	46 24 79 16 76	14 60 25 51 01
28 50 16 43 36	28 97 85 58 99	67 22 52 76 23	24 70 36 54 54	59 28 61 71 96
63 29 62 66 50	02 63 45 52 38	67 63 47 54 75	83 24 78 43 20	92 63 13 47 48
45 65 58 26 51	76 96 59 38 72	86 57 45 71 46	44 67 76 14 55	44 88 01 62 12
39 65 36 63 70	77 45 85 50 51	74 13 39 35 22	30 53 36 02 95	49 34 88 73 61
73 71 98 16 04	29 18 94 51 23	76 51 94 84 86	79 93 96 38 63	08 58 25 58 94
72 20 56 20 11	72 65 71 08 86	79 57 95 13 91	97 48 72 66 48	09 71 17 24 89
75 17 26 99 76	89 37 20 70 01	77 31 61 95 46	26 97 05 73 51	53 33 18 72 87
37 48 60 82 29	81 30 15 39 14	48 38 75 93 29	06 87 37 78 48	45 56 00 84 47
68 08 02 80 72	83 71 46 30 49	89 17 95 88 29	02 39 56 03 46	97 74 06 56 17
14 23 98 61 67	70 52 85 01 50	01 84 02 78 43	10 62 98 19 41	18 83 99 47 99
49 08 96 21 44	25 27 99 41 28	07 41 08 34 66	19 42 74 39 91	41 96 53 78 72
78 37 06 08 43	63 61 62 42 29	39 68 95 10 96	09 24 23 00 62	56 12 80 73 16
37 21 34 17 68	68 96 83 23 56	32 84 60 15 31	44 73 67 34 77	91 15 79 74 58
14 29 09 34 04	87 83 07 55 07	76 58 30 83 64	87 29 25 58 84	86 50 60 00 25
58 43 28 06 36	49 52 83 51 14	47 56 91 29 34	05 87 31 06 95	12 45 57 09 09
10 43 67 29 70	80 62 80 03 42	10 80 21 38 84	90 56 35 03 09	43 12 74 49 14
44 38 88 39 54	86 97 37 44 22	00 95 01 31 76	17 16 29 56 63	38 78 94 49 81
90 69 59 19 51	85 39 52 85 13	07 28 37 07 61	11 16 36 27 03	78 86 72 04 95
41 47 10 25 62	97 05 31 03 61	20 26 36 31 62	68 69 86 95 44	84 95 48 46 45
91 94 14 63 19	75 89 11 47 11	31 56 34 19 09	79 57 92 36 59	14 93 87 81 40
80 06 54 18 66	09 18 94 06 19	98 40 07 17 81	22 45 44 84 11	24 62 20 42 31
67 72 77 63 48	84 08 31 55 58	24 33 45 77 58	80 45 67 93 82	75 70 16 08 24
59 40 24 13 27	79 26 88 86 30	01 31 60 10 39	53 58 47 70 93	85 81 56 39 38
05 90 35 89 95	01 61 16 96 94	50 78 13 69 36	37 68 53 37 31	71 26 35 03 71
44 43 80 69 98	46 68 05 14 82	90 78 50 05 62	77 79 13 57 44	59 60 10 39 66
61 81 31 96 82	00 57 25 60 59	46 72 60 18 77	55 66 12 62 11	08 99 55 64 57
42 88 07 10 05	24 98 65 63 21	47 21 61 88 32	27 80 30 21 60	10 92 35 36 12
77 94 30 05 39	28 10 99 00 27	12 73 73 99 12	49 99 57 94 82	96 88 57 17 91
78 83 19 76 16	94 11 68 84 26	23 54 20 86 85	23 86 66 99 07	36 37 34 92 09
87 76 59 61 81	43 63 64 61 61	65 76 36 95 90	18 48 27 45 68	27 23 65 30 72
91 43 05 96 47	55 78 99 95 24	37 55 85 78 78	01 48 41 19 10	35 19 54 07 73
84 97 77 72 73	09 62 06 65 72	87 12 49 03 60	41 15 20 76 27	50 47 02 29 16
87 41 60 76 83	44 88 96 07 80	83 05 83 38 96	73 70 66 81 90	30 56 10 48 59

Table is taken from Table XXXIII of Fisher & Yates': Statistical Tables for Biological, Agricultural and Medical Research published by Longman Group UK Ltd. London (previously published by Oliver and Boyd Ltd, Edinburgh) and by permission of the authors and publishers.

Table A-6. F Distribution

$\alpha = 0.01$

v_1 \ v_2	1	2	3	4	5	6	7	8	9	10	12	15	20	24	30	40	60	120	∞
1	4052.2	4999.5	5403.3	5624.6	5763.7	5859.0	5928.3	5981.6	6022.5	6055.8	6106.3	6157.3	6208.7	6234.6	6260.7	6286.8	6313.0	6339.4	6366.0
2	98.503	99.000	99.166	99.249	99.299	99.332	99.356	99.374	99.388	99.399	99.416	99.432	99.449	99.458	99.466	99.474	99.483	99.491	99.501
3	34.116	30.817	29.457	28.710	28.237	27.911	27.672	27.489	27.345	27.229	27.052	26.872	26.690	26.598	26.505	26.411	26.316	26.221	26.125
4	21.198	18.000	16.694	15.977	15.522	15.207	14.976	14.799	14.659	14.546	14.374	14.198	14.020	13.929	13.838	13.745	13.652	13.558	13.463
5	16.258	13.274	12.060	11.392	10.967	10.672	10.456	10.289	10.158	10.051	9.8883	9.7222	9.5527	9.4665	9.3793	9.2912	9.2020	9.1118	9.0204
6	13.745	10.925	9.7795	9.1483	8.7459	8.4661	8.2600	8.1016	7.9761	7.8741	7.7183	7.5590	7.3958	7.3127	7.2285	7.1432	7.0568	6.9690	6.8801
7	12.246	9.5466	8.4513	7.8467	7.4604	7.1914	6.9928	6.8401	6.7188	6.6201	6.4691	6.3143	6.1554	6.0743	5.9921	5.9084	5.8236	5.7372	5.6495
8	11.259	8.6491	7.5910	7.0060	6.6318	6.3707	6.1776	6.0289	5.9106	5.8143	5.6668	5.5151	5.3591	5.2793	5.1981	5.1156	5.0316	4.9460	4.8588
9	10.561	8.0215	6.9919	6.4221	6.0569	5.8018	5.6129	5.4671	5.3511	5.2565	5.1114	4.9621	4.8080	4.7290	4.6486	4.5667	4.4831	4.3978	4.3105
10	10.044	7.5594	6.5523	5.9943	5.6363	5.3858	5.2001	5.0567	4.9424	4.8492	4.7059	4.5582	4.4054	4.3269	4.2469	4.1653	4.0819	3.9965	3.9090
11	9.6460	7.2057	6.2167	5.6683	5.3160	5.0692	4.8861	4.7445	4.6315	4.5393	4.3974	4.2509	4.0990	4.0209	3.9411	3.8596	3.7761	3.6904	3.6025
12	9.3302	6.9266	5.9526	5.4119	5.0643	4.8206	4.6395	4.4994	4.3875	4.2961	4.1553	4.0096	3.8584	3.7805	3.7008	3.6192	3.5355	3.4494	3.3608
13	9.0738	6.7010	5.7394	5.2053	4.8616	4.6204	4.4410	4.3021	4.1911	4.1003	3.9603	3.8154	3.6646	3.5868	3.5070	3.4253	3.3413	3.2548	3.1654
14	8.8616	6.5149	5.5639	5.0354	4.6950	4.4558	4.2779	4.1399	4.0297	3.9394	3.8001	3.6557	3.5052	3.4274	3.3476	3.2656	3.1813	3.0942	3.0040
15	8.6831	6.3589	5.4170	4.8932	4.5556	4.3183	4.1415	4.0045	3.8948	3.8049	3.6662	3.5222	3.3719	3.2940	3.2141	3.1319	3.0471	2.9595	2.8684
16	8.5310	6.2262	5.2922	4.7726	4.4374	4.2016	4.0259	3.8896	3.7804	3.6909	3.5527	3.4089	3.2588	3.1808	3.1007	3.0182	2.9330	2.8447	2.7528
17	8.3997	6.1121	5.1850	4.6690	4.3359	4.1015	3.9267	3.7910	3.6822	3.5931	3.4552	3.3117	3.1615	3.0835	3.0032	2.9205	2.8348	2.7459	2.6530
18	8.2854	6.0129	5.0919	4.5790	4.2479	4.0146	3.8406	3.7054	3.5971	3.5082	3.3706	3.2273	3.0771	2.9990	2.9185	2.8354	2.7493	2.6597	2.5660
19	8.1850	5.9259	5.0103	4.5003	4.1708	3.9386	3.7653	3.6305	3.5225	3.4338	3.2965	3.1533	3.0031	2.9249	2.8442	2.7608	2.6742	2.5839	2.4893
20	8.0960	5.8489	4.9382	4.4307	4.1027	3.8714	3.6987	3.5644	3.4567	3.3682	3.2311	3.0880	2.9377	2.8594	2.7785	2.6947	2.6077	2.5168	2.4212
21	8.0166	5.7804	4.8740	4.3688	4.0421	3.8117	3.6396	3.5056	3.3981	3.3098	3.1729	3.0299	2.8796	2.8011	2.7200	2.6359	2.5484	2.4568	2.3603
22	7.9454	5.7190	4.8166	4.3134	3.9880	3.7583	3.5867	3.4530	3.3458	3.2576	3.1209	2.9780	2.8274	2.7488	2.6675	2.5831	2.4951	2.4029	2.3055
23	7.8811	5.6637	4.7649	4.2635	3.9392	3.7102	3.5390	3.4057	3.2986	3.2106	3.0740	2.9311	2.7805	2.7017	2.6202	2.5355	2.4471	2.3542	2.2559
24	7.8229	5.6136	4.7181	4.2184	3.8951	3.6667	3.4959	3.3629	3.2560	3.1681	3.0316	2.8887	2.7380	2.6591	2.5773	2.4923	2.4035	2.3099	2.2107
25	7.7698	5.5680	4.6755	4.1774	3.8550	3.6272	3.4568	3.3239	3.2172	3.1294	2.9931	2.8502	2.6993	2.6203	2.5383	2.4530	2.3637	2.2695	2.1694
26	7.7213	5.5263	4.6366	4.1400	3.8183	3.5911	3.4210	3.2884	3.1818	3.0941	2.9579	2.8150	2.6640	2.5848	2.5026	2.4170	2.3273	2.2325	2.1315
27	7.6767	5.4881	4.6009	4.1056	3.7848	3.5580	3.3882	3.2558	3.1494	3.0618	2.9256	2.7827	2.6316	2.5522	2.4699	2.3840	2.2938	2.1984	2.0965
28	7.6356	5.4529	4.5681	4.0740	3.7539	3.5276	3.3581	3.2259	3.1195	3.0320	2.8959	2.7530	2.6017	2.5223	2.4397	2.3535	2.2629	2.1670	2.0642
29	7.5976	5.4205	4.5378	4.0449	3.7254	3.4995	3.3302	3.1982	3.0920	3.0045	2.8685	2.7256	2.5742	2.4946	2.4118	2.3253	2.2344	2.1378	2.0342
30	7.5625	5.3904	4.5097	4.0179	3.6990	3.4735	3.3045	3.1726	3.0665	2.9791	2.8431	2.7002	2.5487	2.4689	2.3860	2.2992	2.2079	2.1107	2.0062
40	7.3141	5.1785	4.3126	3.8283	3.5138	3.2910	3.1238	2.9930	2.8876	2.8005	2.6648	2.5216	2.3689	2.2880	2.2034	2.1142	2.0194	1.9172	1.8047
60	7.0771	4.9774	4.1259	3.6491	3.3389	3.1187	2.9530	2.8233	2.7185	2.6318	2.4961	2.3523	2.1978	2.1154	2.0285	1.9360	1.8363	1.7263	1.6006
120	6.8510	4.7865	3.9493	3.4796	3.1735	2.9559	2.7918	2.6629	2.5586	2.4721	2.3363	2.1915	2.0346	1.9500	1.8600	1.7628	1.6557	1.5330	1.3805
∞	6.6349	4.6052	3.7816	3.3192	3.0173	2.8020	2.6393	2.5113	2.4073	2.3209	2.1848	2.0385	1.8783	1.7908	1.6964	1.5923	1.4730	1.3246	1.0000

Credit: E. S. Pearson, "Tables of Percentage Points of the Inverted Beta (F) Distribution," *Biometrika* 32 (1943): 73–88.

(continued)

Table A-6. *(continued)*

α = 0.05   α = 0.05

v_2\\v_1	1	2	3	4	5	6	7	8	9	10	12	15	20	24	30	40	60	120	∞
1	161.45	199.50	215.71	224.58	230.16	233.99	236.77	238.88	240.54	241.88	243.91	245.95	248.01	249.05	250.09	251.14	252.20	253.25	254.32
2	18.513	19.000	19.164	19.247	19.296	19.330	19.353	19.371	19.385	19.396	19.413	19.429	19.446	19.454	19.462	19.471	19.479	19.487	19.496
3	10.128	9.5521	9.2766	9.1172	9.0135	8.9406	8.8868	8.8452	8.8123	8.7855	8.7446	8.7029	8.6602	8.6385	8.6166	8.5944	8.5720	8.5494	8.5265
4	7.7086	6.9443	6.5914	6.3883	6.2560	6.1631	6.0942	6.0410	5.9988	5.9644	5.9117	5.8578	5.8025	5.7744	5.7459	5.7170	5.6878	5.6581	5.6281
5	6.6079	5.7861	5.4095	5.1922	5.0503	4.9503	4.8759	4.8183	4.7725	4.7351	4.6777	4.6188	4.5581	4.5272	4.4957	4.4638	4.4314	4.3984	4.3650
6	5.9874	5.1433	4.7571	4.5337	4.3874	4.2839	4.2066	4.1468	4.0990	4.0600	3.9999	3.9381	3.8742	3.8415	3.8082	3.7743	3.7398	3.7047	3.6688
7	5.5914	4.7374	4.3468	4.1203	3.9715	3.8660	3.7870	3.7257	3.6767	3.6365	3.5747	3.5108	3.4445	3.4105	3.3758	3.3404	3.3043	3.2674	3.2298
8	5.3177	4.4590	4.0662	3.8378	3.6875	3.5806	3.5005	3.4381	3.3881	3.3472	3.2840	3.2184	3.1503	3.1152	3.0794	3.0428	3.0053	2.9669	2.9276
9	5.1174	4.2565	3.8626	3.6331	3.4817	3.3738	3.2927	3.2296	3.1789	3.1373	3.0729	3.0061	2.9365	2.9005	2.8637	2.8259	2.7872	2.7475	2.7067
10	4.9646	4.1028	3.7083	3.4780	3.3258	3.2172	3.1355	3.0717	3.0204	2.9782	2.9130	2.8450	2.7740	2.7372	2.6996	2.6609	2.6211	2.5801	2.5379
11	4.8443	3.9823	3.5874	3.3567	3.2039	3.0946	3.0123	2.9480	2.8962	2.8536	2.7876	2.7186	2.6464	2.6090	2.5705	2.5309	2.4901	2.4480	2.4045
12	4.7472	3.8853	3.4903	3.2592	3.1059	2.9961	2.9134	2.8486	2.7964	2.7534	2.6866	2.6169	2.5436	2.5055	2.4663	2.4259	2.3842	2.3410	2.2962
13	4.6672	3.8056	3.4105	3.1791	3.0254	2.9153	2.8321	2.7669	2.7144	2.6710	2.6037	2.5331	2.4589	2.4202	2.3803	2.3392	2.2966	2.2524	2.2064
14	4.6001	3.7389	3.3439	3.1122	2.9582	2.8477	2.7642	2.6987	2.6458	2.6021	2.5342	2.4630	2.3879	2.3487	2.3082	2.2664	2.2230	2.1778	2.1307
15	4.5431	3.6823	3.2874	3.0556	2.9013	2.7905	2.7066	2.6408	2.5876	2.5437	2.4753	2.4035	2.3275	2.2878	2.2468	2.2043	2.1601	2.1141	2.0658
16	4.4940	3.6337	3.2389	3.0069	2.8524	2.7413	2.6572	2.5911	2.5377	2.4935	2.4247	2.3522	2.2756	2.2354	2.1938	2.1507	2.1058	2.0589	2.0096
17	4.4513	3.5915	3.1968	2.9647	2.8100	2.6987	2.6143	2.5480	2.4943	2.4499	2.3807	2.3077	2.2304	2.1898	2.1477	2.1040	2.0584	2.0107	1.9604
18	4.4139	3.5546	3.1599	2.9277	2.7729	2.6613	2.5767	2.5102	2.4563	2.4117	2.3421	2.2686	2.1906	2.1497	2.1071	2.0629	2.0166	1.9681	1.9168
19	4.3808	3.5219	3.1274	2.8951	2.7401	2.6283	2.5435	2.4768	2.4227	2.3779	2.3080	2.2341	2.1555	2.1141	2.0712	2.0264	1.9796	1.9302	1.8780
20	4.3513	3.4928	3.0984	2.8661	2.7109	2.5990	2.5140	2.4471	2.3928	2.3479	2.2776	2.2033	2.1242	2.0825	2.0391	1.9938	1.9464	1.8963	1.8432
21	4.3248	3.4668	3.0725	2.8401	2.6848	2.5727	2.4876	2.4205	2.3661	2.3210	2.2504	2.1757	2.0960	2.0540	2.0102	1.9645	1.9165	1.8657	1.8117
22	4.3009	3.4434	3.0491	2.8167	2.6613	2.5491	2.4638	2.3965	2.3419	2.2967	2.2258	2.1508	2.0707	2.0283	1.9842	1.9380	1.8895	1.8380	1.7831
23	4.2793	3.4221	3.0280	2.7955	2.6400	2.5277	2.4422	2.3748	2.3201	2.2747	2.2036	2.1282	2.0476	2.0050	1.9605	1.9139	1.8649	1.8128	1.7570
24	4.2597	3.4028	3.0088	2.7763	2.6207	2.5082	2.4226	2.3551	2.3002	2.2547	2.1834	2.1077	2.0267	1.9838	1.9390	1.8920	1.8424	1.7897	1.7331
25	4.2417	3.3852	2.9912	2.7587	2.6030	2.4904	2.4047	2.3371	2.2821	2.2365	2.1649	2.0889	2.0075	1.9643	1.9192	1.8718	1.8217	1.7684	1.7110
26	4.2252	3.3690	2.9751	2.7426	2.5868	2.4741	2.3883	2.3205	2.2655	2.2197	2.1479	2.0716	1.9898	1.9464	1.9010	1.8533	1.8027	1.7488	1.6906
27	4.2100	3.3541	2.9604	2.7278	2.5719	2.4591	2.3732	2.3053	2.2501	2.2043	2.1323	2.0558	1.9736	1.9299	1.8842	1.8361	1.7851	1.7307	1.6717
28	4.1960	3.3404	2.9467	2.7141	2.5581	2.4453	2.3593	2.2913	2.2360	2.1900	2.1179	2.0411	1.9586	1.9147	1.8687	1.8203	1.7689	1.7138	1.6541
29	4.1830	3.3277	2.9340	2.7014	2.5454	2.4324	2.3463	2.2782	2.2229	2.1768	2.1045	2.0275	1.9446	1.9005	1.8543	1.8055	1.7537	1.6981	1.6377
30	4.1709	3.3158	2.9223	2.6896	2.5336	2.4205	2.3343	2.2662	2.2107	2.1646	2.0921	2.0148	1.9317	1.8874	1.8409	1.7918	1.7396	1.6835	1.6223
40	4.0848	3.2317	2.8387	2.6060	2.4495	2.3359	2.2490	2.1802	2.1240	2.0772	2.0035	1.9245	1.8389	1.7929	1.7444	1.6928	1.6373	1.5766	1.5089
60	4.0012	3.1504	2.7581	2.5252	2.3683	2.2540	2.1665	2.0970	2.0401	1.9926	1.9174	1.8364	1.7480	1.7001	1.6491	1.5943	1.5343	1.4673	1.3893
120	3.9201	3.0718	2.6802	2.4472	2.2900	2.1750	2.0867	2.0164	1.9588	1.9105	1.8337	1.7505	1.6587	1.6084	1.5543	1.4952	1.4290	1.3519	1.2539
∞	3.8415	2.9957	2.6049	2.3719	2.2141	2.0986	2.0096	1.9384	1.8799	1.8307	1.7522	1.6664	1.5705	1.5173	1.4591	1.3940	1.3180	1.2214	1.0000

Table A-7. Duncan's Multiple Ranges

The numbers given in this table are the values of Q_p used to find $R_p = Q_p s_{\bar{x}}$. The value of R_p, then, is the shortest significant range for comparing the largest and smallest of p means arranged in order of magnitude.

(5% LEVEL)

| Degrees of Freedom | \multicolumn{12}{c}{p = number of means within range being tested} | | | | | | | | | | | |
	2	3	4	5	6	7	8	9	10	20	50	100
1	18.00	18.00	18.00	18.00	18.00	18.00	18.00	18.00	18.00	18.00	18.00	18.00
2	6.08	6.08	6.08	6.08	6.08	6.08	6.08	6.08	6.08	6.08	6.08	6.08
3	4.50	4.52	4.52	4.52	4.52	4.52	4.52	4.52	4.52	4.52	4.52	4.52
4	3.93	4.01	4.03	4.03	4.03	4.03	4.03	4.03	4.03	4 03	4.03	4.03
5	3.64	3.75	3.80	3.81	3.81	3.81	3.81	3.81	3.81	3.81	3.81	3.81
6	3.46	3.59	3.65	3.68	3.69	3.70	3.70	3.70	3.70	3.70	3.70	3.70
7	3.34	3.48	3.55	3.59	3.61	3.62	3.63	3.63	3.63	3.63	3.63	3.63
8	3.26	3.40	3.48	3.52	3.55	3.57	3.58	3.58	3.58	3.58	3.58	3.58
9	3.20	3.34	3.42	3.47	3.50	3.52	3.54	3.54	3.55	3.55	3.55	3.55
10	3.15	3.29	3.38	3.43	3.46	3.49	3.50	3.52	3.52	3.53	3.53	3.53
11	3.11	3 26	3.34	3.40	3.44	3.46	3.48	3.49	3.50	3.51	3.51	3.51
12	3.08	3.22	3.31	3.37	3.41	3.44	3.46	3.47	3.48	3.50	3.50	3.50
13	3.06	3.20	3.29	3.35	3.39	3.42	3.44	3.46	3.47	3.49	3.49	3.49
14	3.03	3.18	3.27	3.33	3.37	3.40	3.43	3.44	3.46	3.48	3.48	3.48
15	3.01	3.16	3.25	3.31	3.36	3.39	3.41	3.43	3.45	3.48	3.48	3.48
16	3.00	3.14	3.24	3.30	3.34	3.38	3.40	3.42	3.44	3.48	3.48	3.48
17	2.98	3.13	3.22	3.28	3.33	3.37	3.39	3.41	3.43	3.48	3.48	3.48
18	2.97	3.12	3.21	3.27	3.32	3.36	3.38	3.40	3.42	3.47	3.47	3.47
19	2.96	3.11	3.20	3.26	3.31	3.35	3.38	3.40	3.42	3.47	3.47	3.47
20	2.95	3.10	3.19	3.26	3.30	3.34	3.37	3.39	3.41	3.47	3.47	3.47
30	2.89	3.04	3.13	3.20	3.25	3.29	3.32	3.35	3.37	3.47	3.49	3.49
40	2.86	3.01	3.10	3.17	3.22	3.27	3.30	3.33	3.35	3.47	3.50	3.50
60	2.83	2.98	3.07	3.14	3.20	3.24	3.28	3.31	3.33	3.47	3.54	3.54
120	2.80	2.95	3.04	3.12	3.17	3.22	3.25	3.29	3.31	3.47	3.58	3.60
∞	2.77	2.92	3.02	3.09	3.15	3.19	3.23	3.26	3.29	3.47	3.64	3.74

Credit: D. B. Duncan, *Biometrics* 11 (1955): 1–42 and modified by H. L. Harter, *Biometrics* 16 (1960): 671–685 and *Biometrics* 17 (1961): 321–324.

(continued)

Table A-7. *(continued)*

(1% LEVEL)

Degrees of Freedom	p = number of means within range being tested											
	2	3	4	5	6	7	8	9	10	20	50	100
1	90.00	90.00	90.00	90.00	90.00	90.00	90.00	90.00	90.00	90.00	90.00	90.00
2	14.00	14.00	14.00	14.00	14.00	14.00	14.00	14.00	14.00	14.00	14.00	14.00
3	8.26	8.32	8.32	8.32	8.32	8.32	8.32	8.32	8.32	8.32	8.32	8.32
4	6.51	6.68	6.74	6.76	6.76	6.76	6.76	6.76	6.76	6.76	6.76	6.76
5	5.70	5.89	6.00	6.04	6.06	6.07	6.07	6.07	6.07	6.07	6.07	6.07
6	5.25	5.44	5.55	5.61	5.66	5.68	5.69	5.70	5.70	5.70	5.70	5.70
7	4.95	5.14	5.26	5.33	5.38	5.42	5.44	5.45	5.46	5.47	5.47	5.47
8	4.75	4.94	5.06	5.14	5.19	5.23	5.26	5.28	5.29	5.32	5.32	5.32
9	4.60	4.79	4.91	4.99	5.04	5.09	5.12	5.14	5.16	5.21	5.21	5.21
10	4.48	4.67	4.79	4.87	4.93	4.98	5.01	5.04	5.06	5.12	5.12	5.12
11	4.39	4.58	4.70	4.78	4.84	4.89	4.92	4.95	4.98	5.06	5.06	5.06
12	4.32	4.50	4.62	4.71	4.77	4.82	4.85	4.88	4.91	5.01	5.01	5.01
13	4.26	4.44	4.56	4.64	4.71	4.76	4.79	4.82	4.85	4.96	4.97	4.97
14	4.21	4.39	4.51	4.59	4.65	4.70	4.74	4.78	4.80	4.92	4.94	4.94
15	4.17	4.35	4.46	4.55	4.61	4.66	4.70	4.73	4.76	4.89	4.91	4.91
16	4.13	4.31	4.42	4.51	4.57	4.62	4.66	4.70	4.72	4.86	4.89	4.89
17	4.10	4.28	4.39	4.48	4.54	4.59	4.63	4.66	4.69	4.83	4.87	4.87
18	4.07	4.25	4.36	4.44	4.51	4.56	4.60	4.64	4.66	4.81	4.86	4.86
19	4.05	4.22	4.34	4.42	4.48	4.53	4.58	4.61	4.64	4.79	4.84	4.84
20	4.02	4.20	4.31	4.40	4.46	4.51	4.55	4.59	4.62	4.77	4.83	4.83
30	3.89	4.06	4.17	4.25	4.31	4.37	4.41	4.44	4.48	4 65	4.77	4.78
40	3.82	3.99	4.10	4.18	4.24	4.30	4.34	4.38	4.41	4.59	4.74	4.76
60	3.76	3.92	4.03	4.11	4.17	4.23	4.27	4.31	4.34	4.53	4.71	4.76
120	3.70	3.86	3.96	4.04	4.11	4.16	4.20	4.24	4.27	4.47	4.67	4.77
∞	3.64	3.80	3.90	3.98	4.04	4.09	4.14	4.17	4.20	4.41	4.64	4.78

Table A-8. Factors for Computing Control Chart Limits

Observations in sample, n	For averages A_2	For R chart D_3	For R chart D_4
2	1.880	0	3.267
3	1.023	0	2.575
4	0.729	0	2.282
5	0.577	0	2.115
6	0.483	0	2.004
7	0.419	0.076	1.924
8	0.373	0.136	1.864
9	0.337	0.184	1.816
10	0.308	0.223	1.777
11	0.285	0.256	1.744
12	0.266	0.284	1.716
13	0.249	0.308	1.692
14	0.235	0.329	1.671
15	0.223	0.348	1.652
16	0.212	0.364	1.636
17	0.203	0.379	1.621
18	0.194	0.392	1.608
19	0.187	0.404	1.596
20	0.180	0.414	1.586

Table A-9.   Military Standard 105E Sample Size Code Letters

Lot or batch size			Special inspection levels— Used for destructive tests or where sample sizes must be small				General inspection levels— Confidence levels		
							90%	95%	99%
			S-1	S-2	S-3	S-4	I	II	III
2	to	8	A	A	A	A	A	A	B
9	to	15	A	A	A	A	A	B	C
16	to	25	A	A	B	B	B	C	D
26	to	50	A	B	B	C	C	D	E
51	to	90	B	B	C	C	C	E	F
91	to	150	B	B	C	D	D	F	G
151	to	280	B	C	D	E	E	G	H
281	to	500	B	C	D	E	F	H	J
501	to	1200	C	C	E	F	G	J	K
1201	to	3200	C	D	E	G	H	K	L
3201	to	10000	C	D	F	G	J	L	M
10001	to	35000	C	D	F	H	K	M	N
35001	to	150000	D	E	G	J	L	N	P
150001	to	500000	D	E	G	J	M	P	Q
500001	and	over	D	E	H	K	N	Q	R

Table A-10. Master Tablbe for Normal Inspection (Single Sampling)—MIL-STD-105E

Acceptable quality levels (normal inspection)

Sample size code letter	Sample size	0.010 Ac Re	0.015 Ac Re	0.025 Ac Re	0.040 Ac Re	0.065 Ac Re	0.10 Ac Re	0.15 Ac Re	0.25 Ac Re	0.40 Ac Re	0.65 Ac Re	1.0 Ac Re	1.5 Ac Re	2.5 Ac Re	4.0 Ac Re	6.5 Ac Re	10 Ac Re	15 Ac Re	25 Ac Re	40 Ac Re	65 Ac Re	100 Ac Re	150 Ac Re	250 Ac Re	400 Ac Re	650 Ac Re	1,000 Ac Re
A	2															0 1			1 2	2 3	3 4	5 6	7 8	10 11	14 15	21 22	30 31
B	3												0 1					1 2	2 3	3 4	5 6	7 8	10 11	14 15	21 22	30 31	44 45
C	5										0 1						1 2	2 3	3 4	5 6	7 8	10 11	14 15	21 22	30 31	44 45	
D	8								0 1				1 2	2 3	3 4	5 6	7 8	10 11	14 15	21 22							
E	13							0 1				1 2	2 3	3 4	5 6	7 8	10 11	14 15	21 22								
F	20						0 1				1 2	2 3	3 4	5 6	7 8	10 11	14 15	21 22									
G	32					0 1				1 2	2 3	3 4	5 6	7 8	10 11	14 15	21 22										
H	50				0 1				1 2	2 3	3 4	5 6	7 8	10 11	14 15	21 22											
J	80			0 1				1 2	2 3	3 4	5 6	7 8	10 11	14 15	21 22												
K	125		0 1				1 2	2 3	3 4	5 6	7 8	10 11	14 15	21 22													
L	200	0 1				1 2	2 3	3 4	5 6	7 8	10 11	14 15	21 22														
M	315				1 2	2 3	3 4	5 6	7 8	10 11	14 15	21 22															
N	500			1 2	2 3	3 4	5 6	7 8	10 11	14 15	21 22																
P	800		1 2	2 3	3 4	5 6	7 8	10 11	14 15	21 22																	
Q	1,250																										
R	2,000	1 2	2 3	3 4	5 6	7 8	10 11	14 15	21 22																		

↓ = use first sampling plan below arrow. If sample size equals, or exceeds, lot or batch size, do 100 % inspection.

↑ = use first sampling plan above arrow.

Ac = acceptance number.

Re = rejection number.

Table A-11. Percentage Points of the Studentized Range, $q = (x_n - x_1)/S_v$

LOWER 5% POINTS

v \ n	2	3	4	5	6	7	8	9	10
1	18.00	27.00	32.80	37.10	40.40	43.10	45.40	47.10	49.10
2	6.09	8.30	9.80	10.90	11.70	12.40	13.00	13.50	14.00
3	4.50	5.91	6.82	7.50	8.04	8.48	8.85	9.18	9.46
4	3.93	5.04	5.76	6.29	6.71	7.05	7.35	7.60	7.83
5	3.64	4.60	5.22	5.67	6.03	6.33	6.58	6.80	6.99
6	3.46	4.34	4.90	5.31	5.63	5.89	6.12	6.32	6.49
7	3.34	4.16	4.68	5.06	5.36	5.61	5.82	6.00	6.16
8	3.26	4.04	4.53	4.89	5.17	5.40	5.60	5.77	5.92
9	3.20	3.95	4.42	4.76	5.02	5.24	5.43	5.60	5.74
10	3.15	3.88	4.33	4.65	4.91	5.12	5.30	5.46	5.60
11	3.11	3.82	4.26	4.57	4.82	5.03	5.20	5.35	5.49
12	3.08	3.77	4.20	4.51	4.75	4.95	5.12	5.27	5.40
13	3.06	3.73	4.15	4.45	4.69	4.88	5.05	5.19	5.32
14	3.03	3.70	4.11	4.41	4.64	4.83	4.99	5.13	5.25
15	3.01	3.67	4.08	4.37	4.60	4.78	4.94	5.08	5.20
20	2.95	3.58	3.96	4.23	4.45	4.62	4.77	4.90	5.01
40	2.86	3.44	3.79	4.04	4.23	4.39	4.52	4.63	4.74
60	2.83	3.40	3.74	3.98	4.16	4.31	4.44	4.55	4.65
120	2.80	3.36	3.69	3.92	4.10	4.24	4.36	4.48	4.56
inf	2.77	3.31	3.68	3.86	4.03	4.17	4.29	4.39	4.47

Table A-11. *(continued)*

UPPER 5% POINTS

n	2	4	6	8	10	12	14	16	18
>10	0.09	0.75	1.20	1.52	1.74	1.91	2.05	2.17	2.26
11	0.09	0.75	1.21	1.52	1.75	1.92	2.07	2.18	2.28
12	0.09	0.75	1.21	1.53	1.76	1.93	2.08	2.20	2.30
13	0.09	0.75	1.22	1.53	1.76	1.94	2.09	2.21	2.31
14.	0.09	0.75	1.22	1.54	1.77	1.95	2.10	2.22	2.33
15	0.09	0.75	1.22	1.54	1.77	1.95	2.11	2.23	2.34
16	0.09	0.75	1.22	1.54	1.78	1.96	2.11	2.24	2.34
17	0.09	0.75	1.22	1.55	1.78	1.97	2.12	2.25	2.35
18	0.09	0.75	1.22	1.55	1.79	1.97	2.12	2.25	2.36
19	0.09	0.75	1.23	1.55	1.79	1.98	2.13	2.26	2.37
20	0.09	0.75	1.23	1.55	1.79	1.98	2.13	2.27	2.37
40	0.09	0.76	1.24	1.57	1.82	2.02	2.18	2.32	2.43
inf	0.09	0.76	1.25	1.60	1.86	2.07	2.24	2.39	2.52

Table A-12.   Table of α (Producer's Risk at 5%)

df \ $k=$ 2	3	4	5	6	8	10	15	20	30
6	1.73	2.59	2.94	3.19	3.37					
8	1.63	2.39	2.71	2.92	3.09	3.33				
10	1.58	2.29	2.58	2.78	2.93	3.15	3.31			
15	1.51	2.16	2.42	2.60	2.74	2.93	3.07	3.32		
20	1.48	2.10	2.35	2.52	2.64	2.83	2.96	3.18	3.33	
30	1.44	2.04	2.28	2.44	2.56	2.73	2.86	3.06	3.19	3.37
40	1.43	2.01	2.25	2.40	2.52	2.69	2.80	3.00	3.13	3.29
60	1.41	1.98	2.21	2.36	2.48	2.64	2.76	2.94	3.06	3.22
120	1.40	1.95	2.18	2.33	2.44	2.60	2.71	2.88	3.00	3.15
∞	1.39	1.93	2.15	2.29	2.40	2.55	2.65	2.82	2.94	3.08

k = number of means being compared.

df = degrees of freedom in estimate of error variance, $s_{\bar{x}}^2 = \hat{\sigma}_{\bar{x}}^2$.

Bibliography

Alvarez, R. J. 1994. Managing total quality in foodservice. *Food Technology* 48(9):140–143

American Society for Quality Control. T151, 1986. *Food processing industry quality system guidelines.*

American Supplier Institute. 1990. *Total quality management.* Dearborn, MI: American Supplier Institute.

American Institute of Baking. 1988. *Quality assurance manual for food processors.* Manhattan, KS: AIB.

American National Standards Institute. 1981. *ANSI/ASQC Z1.4, Military Standard 105D, Sampling procedures and tables for inspection by attributes.* New York: American Standards Association.

American Society for Quality Control. 1986. *Food processing industry quality system guidelines.* Milwaukee, WI: ASQC.

American Society for Testing and Materials. 1968. *Basic principles of sensory evaluation.* ASTM Special Technical Publication # 433. Philadelphia, PA: ASTM.

―――. 1977. *Manual on sensory testing methods.* Philadelphia, PA: ASTM.

ASM International (formerly American Society for Metals). *Metals handbook.* 8th ed. Vol. 11. Metals Park, OH: ASM International.

Boelens, and Haring. 1982. Paper presented at the annual meeting of the Institute of Food Technologists.

Box, G., J. S. Hunter, and W. G. Hunter. 1978. *Statistics for experimenters.* New York: Wiley.

Box, G., and N. R. Draper. 1969. *Evolutionary operation.* New York: Wiley.

Braverman, D. 1981. *Fundamentals of statistical quality control.*

Centers for Disease Control. 1981. *Foodborne disease outbreaks.* Atlanta, GA: CDC.

Crosby, P. 1979. *Quality is free.* New York: McGraw-Hill.

Dambolena, I., and A. Rao. 1994. What is six sigma anyway? *Quality* 33(11):10.

DataMyte Corporation. 1987. *DataMyte handbook.* Minnetonka, MN: DataMyte Corporation.

Deming, S. N., and S. L. Morgan. 1988. *Experimental design: A chemometric approach.* New York: Elsevier.

Deming, W. E. 1975. On some statistical aids toward economic production. *Interfaces* 5 (The Institute of Management Sciences) 4.

———. 1982. *Quality, productivity, and competitive position.* Cambridge, MA: MIT Center for Advanced Engineering Study.

———. 1986. *Out of the crisis.* Cambridge, MA: MIT Press.

Dodge, H. F. 1948. Administration of a sampling inspection plan. *Industrial Quality Control* 5(3):12–19.

Feigenbaum, A. V. 1983. *Total quality control, engineering and management.* 3rd. ed. New York: McGraw-Hill.

Food and Drug Administration. 1989. *Code of Federal Regulations 21, Section 101.105.* Washington, DC: Government Printing Office.

Food Processors Institute. 1982. *Practical net weight control.* Washington, DC: FPI.

Ford Motor Company. 1985. *Continuing process control and process capability improvement.* Plymouth, MI: Ford Motor Company.

Freedman, W. 1984. *Products liability.* New York: Van Nostrand Reinhold.

Grant, E. L., and R. S. Leavenworth. 1988. *Statistical quality control.* 6th ed. New York: McGraw-Hill.

Grocery Manufacturers of America. 1974. *Guidelines for product recall.* Washington, DC: GMA.

Hansen, B. 1952. *Quality control theory and applications.* New York: Wiley.

Heckel, J. 1988. Built in quality at Greely. *Quality* 27:20.

Hertz, P. 1987. The Deming method. *Quality,* pp. Q-32-34.

Hradesky, J. L. 1988. *Productivity quality improvement.* New York: McGraw-Hill.

Hubbard, M. R. 1987. New quality control techniques for the food industry. *Hornblower* 39(5): 8–10.

Institute of Food Technologists, Sensory Evaluation Division. 1981. Sensory evaluation guide for testing food. *Food Technology* November: 50.

Ishikawa, K. 1982. *Guide to quality control.* Tokyo: Asian Productivity Center.

———. 1985. *What is total quality control?* Englewood Cliffs, NJ: Prentice-Hall.

Juran, J. M. 1988. *Quality control handbook.* New York: McGraw-Hill.

Juran, J. M., and F. M. Gryna. 1970. *Quality planning and analysis.* 2nd ed. New York: McGraw-Hill.

Kennedy, C. W., and D. E. Andrews. 1987. *Inspection and gaging.* 5th ed. New York: Industrial Press.

McGonnagle, W. J. 1971. *Nondestructive testing.* New York: Gordon and Breech.

Merrill Lynch. 1995. *Global Investment Strategy.* March 21.

Miller, J. C., and J. N. Miller. 1988. *Statistics for analytical chemistry.* New York: Wiley (Halsted).

National Institute of Standards and Technology (formerly National Bureau of Standards). 1988. *Handbook 133.* 3rd ed., with Supplements 1, 2, and 3, dated 1990. 1991, and 1992. *Checking the net contents of packaged goods.* Washington, DC: Government Printing Office.

———. 1995. *Handbook 130, Uniform laws and regulations.* Washington, DC: Government Printing Office.

Ott, E. 1975. *Process and control: Trouble-shooting and interpretation of data.* New York: McGraw-Hill.

Peters, T., and R. H. Waterman, Jr. 1982. *In search of excellence.* New York: Harper & Row.

Pierson, M. D., and D. A. Corlett. 1992. *HACCP principles and applications.* New York: Chapman & Hall.

Quality Progress. 1990. The tools of quality. June through December.

Rickmers, A. D., and H. N. Todd. 1967. *Statistics: An introduction.* New York: McGraw-Hill.

Schilling, E. G. *Statistical quality control.* 6th ed. New York: McGraw-Hill.

Tadikamalla, P. R. 1994. The confusion over six-sigma quality. *Quality Progress* 27(11):83–91.

U.S. Department of Agriculture, Food Safety and Quality Service, 1988. *Code of Federal Regulations 9, Section 317.2.* Washington, DC: Government Printing Office.

U.S. General Accounting Office. 1991. *Management Practices.* GAO/NSIAD-91-190.

University of Iowa ASQC Section 1303. 1965. *Quality control training manual.* 2nd ed. Iowa City, IA: University of Iowa.

———. 1986. *Statistical quality control handbook.* Iowa City, IA: University of Iowa.

Index